# NO SET AGENDA

# NO SET AGENDA

## Australia's Catholic Church Faces an Uncertain Future

### PAUL COLLINS

David Lovell Publishing
Melbourne Australia

Published by
David Lovell Publishing
308 Victoria Street
Brunswick, Victoria 3056
Australia
Telephone (03) 380 6728

First published 1991



Cover design and illustration by David Constable
Cover illustration from a photograph by Michael Coyne
Typeset in 10½/12 California by
   Optima Typesetters, Brunswick East
Printed and bound in Australia by The Book Printer, Victoria

National Library of Australia
Cataloguing-in-Publication data
Collins, Paul, 1940–
   No set agenda: the Catholic Church faces an uncertain future.
   Includes index.
   ISBN 1 86355 020 8.

   1. Catholic Church — Australia. 2. Mission of the church. I. Title.
282.94

# *DEDICATION*

In honour of one of the greatest
of all of us in the year of the bi-centenary of his death

JOHANNES CHRYSOSTOMOS WOLFGANG AMADEUS MOZART
(1756–1791)

# CONTENTS

# *PREFACE*

I begin with an apology for the presumption of dedicating such an ordinary book to a man who was so extraordinary. Few people have contributed more to humankind. In the 200 years since his death, Mozart has profoundly enriched the lives of all who take the trouble to listen to him attentively. I am proud to say that I have been a devotee since I was ten years old, when I somehow came into possession of an old microgroove record of Mozart's *German Dances*, the *Eine kleine Nachtmusik* and some of the opera overtures. The music was played by that fine orchestra first brought together by Toscanini, the Columbia Symphony Orchestra. The Mozart works were conducted by Bruno Walter on a record entitled *In the Gardens of Mirabell*. It was only much later that I learned that the Mirabell Palace was built in the early 17th century by the Prince-Archbishop Wolf Dietrich for his mistress and their children, and that the Mirabell Gardens were laid out in the early 18th century by the great baroque architect, Johann Fischer von Erlach.

In a way, mention of this beautiful part of Salzburg brings us to the topic of this book. In the course of its history the Catholic church has been and done many things, including running an odd ecclesiastical city-state, like Salzburg. The church has been both radical and reactionary, usually both at the same time! In my 1986 book *Mixed Blessings*, I argued that in many places in the contemporary world, Catholicism is an extraordinarily creative and radical force.[1] But in the book I also argued that Australia does not fit into this radical and creative pattern. Most outsiders — and many insiders — see the Australian Catholic church as a seemingly dull and lifeless institution, bereft of leadership, committed to caution and socially conservative. It seems to have no real sense of direction and no clear agenda for the decade leading up to the millenium.

A closer look at the Australian church shows that this judgement is not entirely true. A number of creative things are happening here in Australia. But the problem that I touched on in *Mixed Blessings* still remains. Things tend to happen ad hoc, without any sense of planning and certainly without co-ordination. No one seems to be able to stand back and see the whole picture. This what I have tried to do in this book. But beyond an overview is the task of trying to

---

1. Collins, Paul, *Mixed Blessings: John Paul II and the Church of the Eighties*, Penguin Books, Ringwood, 1986.

set up some agenda items that will focus the energy of the Australian Catholic community as it moves toward the year 2000. So I have also set myself the task in this book of setting up at least a partial agenda for the Australian Catholicism. No doubt this is both an ambitious and presumptuous task.

Since I published *Mixed Blessings* in 1986 I have received letters accusing me of betraying Catholicism; I have been denounced by reactionaries of disloyalty to the pope, been called a 'Judas', a 'subversive' and told that I do not have the 'moral courage' to stand for the truth. There is a common thread running through all these views: they all define the Catholic church far too narrowly. Paradoxically, the appeal of the reactionary is to tradition. 'If only we could recover past tradition and restore orthodoxy, all would be well', they tell us. In the reactionary mind, the perfection of tradition is always found in the period before 1962 and the Second Vatican Council.

The historical irony is that most so-called traditionalists are not all that traditional. They identify tradition with a particular period of church history and somehow that period becomes the norm against which everything is judged. Nowadays reactionaries usually focus on the post-reformation era and specifically on the period of church history since the mid-19th century, up to and including the 1950s.

When I was a seminarian, we had a lecturer who thought he had a strong historical sense. He used to tell us: 'The church has traditionally taught this, or that, or the other thing'. He was often followed by a much more perceptive lecturer who asked: 'But how old is that tradition?' When you traced back the history of a particular 'constant tradition' you found it came into vogue in the post-reformation era, or in the 19th century, and that behind it was what Karl Rahner has called a 'forgotten truth' that was often far more ancient and traditional. So often it is these forgotten truths that Catholics need to rediscover from their tradition.

The vast majority of Australian Catholics belong to what might be loosely called the liberal mainstream. They are happy with the changes in the church that resulted from Vatican Council II, and, like Australians generally, are not much given to ideology. We are an essentially practical people. So there is not a great deal of enthusiasm among ordinary Catholics for those who want to drag the church back to the 19th century. But that does not mean that the Catholic community does not want to understand what is going on in the church. There is a hunger for a *context*, a way of comprehending the variegated religious experiences that all Australian Catholics have had. I hope that this book will help in this process of contextualisation.

At this point I must make a comment about one important issue that is part of this context — the role of women in the church. A recently published book of women's stories carries the perceptive title *A Remarkable Absence of Passion.* [2] There have been times when I have been accused of lacking passion especially towards the question of women's role in the church. I consider the shift in the role of women in society to be the most important and profound social mutation in western culture for many centuries. If the church does not address this issue it will not only lose half its constituency; it will lose its right to articulate the basic issues facing humankind and in the process lose its own soul. This book has no separate section on women. Every time the word 'laity' or 'lay people' is used it includes — of course — women. The issue of women is part of the discussion of every topic in this book.

Australian reactionaries are a very vociferous, though nevertheless tiny minority of Catholics. Far more notice is taken of them than their numbers warrant. They are not really *conservative.* Conservatives value the past in order to maintain a sense of stability and continuity, but, at the same time, maintain a reasonably open perspective on the future. In this sense it would be true to say that most Australian Catholics are reasonably conservative. They do not want to overthrow the established order of the church, but they do want the church to be open to the future. In fact the terms 'liberal' and 'conservative' are increasingly meaningless within the context of the church. It is closer to reality to talk about a large majority of mainstream Catholics and a tiny minority of reactionary Catholics.

Let me tell you my position right at the beginning: I am a conservative and a firm believer in tradition. But tradition, as I will argue later in this book, is a complex theological reality that involves an appreciation of the past, a willingness to live unashamedly in the present, and an imaginative hopefulness that looks to the future. It is within this context that I want to look at the Australian Catholic church. But I need to be honest with you: there is also a sense in which I have now moved beyond the Australian Catholic mainstream. There are opinions expressed in this book which some people will

---

2. McManus, Nora (ed.), *A Remarkable Absence of Passion: Women Address the Catholic Church*, David Lovell Publishing, 1991.

find too radical to stomach. I am not apologising in advance for these views. Unless someone tries to imagine a possible future, there will be no future. In fact, if Australians lack anything, it is imagination. I hope to inject a little of that into our contemporary Catholic debate.

# ❖ *1* ❖

# *WHY MORE ON AUSTRALIAN CATHOLICISM?*

The answer to this question is quite simple: I am both an incurable talker and an incurable Catholic. Because I am a talker I work in the media where I naturally tend to draw criticism. Recently, a friend told me he was having lunch with an Australian Catholic bishop. My name came up in conversation and the bishop commented forcefully that I 'obviously hated the Catholic church'! It was probably one of those off-the-cuff comments that should not be taken too seriously. But the report of the comment came to me just after the secretary of the Australian Catholic Bishop's Conference was quoted in *The Australian* as saying that some of my ideas on the clergy and bishops, in an article I had written in the ecumenical magazine *National Outlook*, were motivated by my 'own inner struggle and insecurity'. I was fascinated by the secretary's psychological insight (he has recently been appointed a bishop), given that I had experienced very little 'inner struggle' for some considerable time. After all, at fifty, you do tend to settle down a bit!

I would agree that there is an 'inner struggle' going on. But it is not for my psyche; it is for the enormously more important reality of the future of the Australian Catholic church. Which is the second reason for this book. My concern in writing is the state of the church. It is in a serious but unarticulated, even unrealised crisis. Many people working in the administration of the church's structures, the people who seem to assume that they 'own' Catholicism in this country, simply cannot see what is happening around them. Their ecclesiastical blinkers prevent them. Similarly, many working hard in ministry cannot see what is happening around them because they either do not have the time, or do not want to lift their heads above the ruck to reflect on contemporary reality.

When I say that the Australian Catholic church is in crisis, I do not claim any special insight. It is simply that lately I have seen things from a different point of view. I have not lived a conventional priestly life for the last decade. (It has been rumoured on a number of occasions that I had 'left' the ministry. But I keep on annoying the rumour-mongers by re-appearing!) I have not had a coherent plan over the last ten years, except that, at most, I wanted to create some 'space' for new things to happen, to discover new ways of seeing things in a church which has increasingly become claustrophobic. What living this way has done is to give me another perspective, another way of seeing the church from the outside while remaining committed.

Over the last three years I have worked with the Australian Broadcasting Corporation, specialising in religion in Australia. This has also given me a unique perspective on Catholicism. In the ABC Religious Department we say that our job involves dealing with a two-way traffic. We firstly reflect the religious communities of Australia to the wider world through radio and television; in other words we try to report the story of church, religion and spiritual experience in this country. But we also reflect back to Australia's religious communities the observations and criticisms of the wider Australian society. We do this by trying to get religious people to address and answer the questions that our culture directs to them and to their traditions.

Catholic bishops and clergy, despite their protestations about openness, are not very used to this. They find it easy enough to deflect the criticisms of those who do not know what they are talking about; it is a lot harder to be confronted by those who do and who, at the same time, maintain a vocational commitment to Catholicism as well.

This is why those who are afraid of comment and questions resort to personal attack, the type of thing that in the old days of formal

Latin disputations in the seminary we used to call the *argumentum ad hominem*, the type of argument that side-stepped the point of the argument to attack the arguer. To criticise the church is not necessarily to hate it. And one who expresses criticism need not be speaking out of some inner struggle or insecurity!

On the contrary, I would suggest that such criticism might be motivated by affection and loyalty, even love. It is very easy to dismiss criticism when you can attribute it to 'inner struggle and insecurity' or even to 'hatred'. It is more difficult to admit that it may be motivated by commitment to the same church as yourself.

Unfortunately, however, I again run the risk of annoying those who identify criticism with hatred. Fundamentally, this book argues that the Australian Catholic church is facing a situation as serious as any that it has faced in its 200 year history. One of the strengths of this church until recently has been that it has been very much part of Australian culture. Historically, this was partly based on the fact that the church was ethnically Irish and socially working class, especially in early part of the 20th century. Within the Catholic community there was a close relationship between clergy and laity, who shared both ethnic and social origins. It was seen as the task of the clergy and bishops to articulate Catholic claims. This affinity also arose from a clear sense of what it was to be Catholic. In fact some critics have argued that Catholicism in Australia had generally become far more 'Australian' than 'Catholic'.

In this book I will argue that the contemporary Australian Catholic church has, to a considerable extent, lost its sense of being both 'Catholic' and 'Australian'. Its leadership has largely retired from engagement with the issues that face our contemporary society and many church people have retired into their own, safe sub-culture. This is in interesting contrast to the wider Catholic church, which in many parts of the world is seriously engaged in the religious, spiritual, ethical, cultural, political, environmental and economic issues of our time, as it struggles to relate its tradition and teaching to contemporary needs. I described some of this engagement in *Mixed Blessings*, where I also described the Australian church as something of a 'backwater' of world Catholicism. But even if it is a backwater, that does not mean that Australian Catholics should not be concerned about it.

This book is an expression of my concern about the Australian church and its seeming increasing irrelevance to Catholics who are, and still want to be, its faithful adherents. It is my conviction that the Catholic tradition has within it enormous resources and an adaptive ability to address the issues that are being raised by fair-minded people in contemporary Australia. The fact that the Catholic tradition is not seriously engaged or even considered when our society debates fundamental questions of meaning, ethics and social morality, is not the fault of the media, the politicians or 'secular humanists'. It is the fault of Catholics themselves, and especially the leadership of the church, both clerical and lay. In this book I will try to analyse why Australian Catholicism is not meeting the challenges and try to suggest ways in which it might break out of its self-enclosure and move toward the future.

## Has Catholism lost its chance?

For the first time in the history of white Australia, the 1986 national census shows that the Catholic church is now the largest religious group in the country. The Catholic community is also younger than any of the other major Christian churches, with just over one million adherents under the age of fourteen and three quarters of a million between fifteen and twenty-four. [1]

But just at the time when Catholics might be tempted to indulge in a little self-congratulation, there are many signs that the Australian church has entered a perilous period in its development. Catholicism is not facing an external crisis, but a loss of confidence and self-identity. The core of the problem is that the Australian church has become lethargic and passionless; it has lost its sense of direction. It has few articulated goals and no strong convictions about what its role should be in contemporary Australian society. It has turned in on itself.

The evidence for this can be seen most clearly in the core group that carries much of the burden of the church's pastoral ministry. As in every institution that depends on its main work being done by a group of people with a strong vocational sense, so in the Catholic church there is a focal group of laity, priests and members of religious orders, an élite for whom participation in church life is the centre of their being and the core of their vocation. They know that they

---

1. In the 1986 census Catholics constituted 26.1% of the population and Anglicans 23.9%. See *Australia in Profile: Census 86*, Australian Bureau of Statistics, p. 19. See also Bentley, Peter, Blombery, 'Tricia, & Hughes, Philip J., *A Yearbook for Australian Churches 1991*, Christian Research Association, Hawthorn, 1990, pp. 156, 158–159.

believe and they also have a sense of why they believe. It is this group that is now most disillusioned in the Australian Catholic church. On a broader scale, many practising Catholics feel leaderless and bereft, as the institutional church seemingly retires into a safe sub-culture. They are disappointed with a clergy who have failed to nurture their spirituality, to support them when they try to participate in ministry and who seem uninterested in the issues that are important to them as lay people.

As a result many have stopped practising as Catholics. Some just drop out, while the more passionate pass through a painful period of loss as they leave active involvement behind. The tragedy is that the church is losing its real lay and priestly leadership cadre, the very people who, because of their faith, sensitivity and conviction, are essential for it to move it into the future.

Take Kevin, for example. From the 1960s onwards he participated fully in the life of the church. He got together a lay group, was involved in the charismatic prayer movement, worked full time for several years teaching spirituality and later was the interior designer of an award-winning Catholic centre. But in the end he felt forced out of the church. As he himself says:

Most of my friends just had to find ministry elsewhere . . . For so many years I fought for everything from married priests to lay spirituality . . . but in the end my wife and I and family just walked away. Probably the present Pope just alienated us more completely.

He told me of a priest friend who had worked for twenty years to support lay people and to build the church both physically and spiritually in one of the more deprived areas of one of the capital cities. I have also heard many other people praise the dedication of this particular priest. Kevin told me that when this priest followed his conscience and left the active priestly ministry,

He had no option but to go and get married in a Protestant church with a congregation full of his friends, nuns, priests and even a bishop all frozen to their seats like bloody statues! No one could do, or did anything.

Kevin himself left the church partly because no one took his attempts to develop a cosmological and ecological spirituality seriously. As I will argue later in the book, this is one of the key issues facing the Australian church. But his story does not end with despair. This is how he describes his spirituality now:

God has a way of providing sacraments in the most imaginative ways . . . What was the probably the most interesting aspect of leaving the Catholic institution was that after a period of 'spiritual nuclear winter', I found a wider, more tolerant community of faith people, loosely connected and roughly called the 'new age' movement. A large percentage of these people are, of course, ex-Catholics'!

Kevin's story illustrates what many of the best Australian Catholics have experienced. Ignored and even denigrated within the church, they have been forced to walk elsewhere. Sometimes, like Kevin, they have found another, enriching vision of God and human life. But not always; priests, for instance, who have left the active ministry have sometimes been deprived of even basic human rights. There have been a number of cases where such men have been denied work in Catholic schools, despite their qualifications; in one situation that eventually went before a state ombudsman, a former priest was denied work as a librarian in a Catholic primary school on the orders of the local bishop. The ombudsman declined to enter the dispute on the grounds of the separation of church and state!

The reason why people abandon the church lies principally with the attitude of the church itself. And Catholics should not fool themselves that their numbers have not dropped. If you take weekly Mass attendance as a norm — an imperfect, but real measure of practical Catholicism — in the period 1960 to the mid-1980s, the rate of Catholic practice has plummetted. (It also has to be said that Anglican and Protestant rates also fell sharply during this period, but weekly church attendance has never been so important for them.) Catholic women have fallen from a sixty per cent weekly attendance rate in 1967 to thirty-six per cent in 1985 and Catholic male attendance has fallen from forty-two per cent in 1967 to thirty-two per cent in 1985. [2] In Melbourne, the annual Mass count (when each parish counts attendances at each parish over three consecutive Sundays in May) shows that attendances have dropped from 218 445 in 1987 to 211 670 in 1990. The Catholic population of the Archdiocese of Melbourne at the 1986 census was 879 000. [3]

Some sociologists have argued that the decline in church attendance

---

2. McCullum, John, 'Secularisation in Australia between 1966 and 1985. A Research Note', *Australian and New Zealand Journal of Sociology*, 23/3(1987), pp. 414–415.
3. The figures are available from the 'Mass Count Report 1990';, Catholic Research Office for Pastoral Planning, Archdiocese of Melbourne.

levelled off in the 1980s. However, the National Social Science Survey indicates that the trend of decline has continued and that the rate of practice has now dropped to about nineteen per cent of all Catholics attending weekly Mass in 1990.[4] The argument that church attendance has levelled off in the 1980s does make some sense when you look at the 'monthly' or 'nearly every week' figures, which show that about thirty-seven per cent of Catholics go to Mass monthly or bi-monthly. In other words, the attitude of these people has shifted from a sense of compulsory weekly attendance to voluntary attendance once or twice a month. They are what 'Tricia Blombery and Philip J. Hughes have called 'principalists'; for them 'Christian values [are] not seen in terms of specific rules, but as ethical principles'.[5]

In spite of all of this, there is one place where the Catholic presence has not dropped off — prison! When asked what percentage of prisoners were Catholic, Melbourne prison chaplain, Father Peter Norden, said that the Catholic percentage was higher than in the general community — about thirty-five per cent of those in jail were Catholics.[6] Back in the 1930s and 1940s when Dr Rumble's *Radio Replies* was first published, bigoted Protestants claimed (correctly) that there were more Catholics in jail than any other section of the community. Never at a loss for an answer, Rumble replied that Christ was more likely to be found in jail because it was there that he was needed most and that he was more likely to consort with sinners than with Protestants!

But, returning to the wider world: clearly a very large group of Australian Catholics has simply stopped going to Mass. For these people that absence is symbolic of the fact that they have also stopped participating in the life of the church.

## Church vs 'world'

Why has this happened? The leadership of the church generally blames this on 'secular humanism' and on the irreligious materialism that is said to be characteristic of our society. Church hierarchs often borrow that slippery word 'secularisation' from sociologists and say that Australian secularism has invaded the Catholic psyche and

---

4. See part sample of the 1989 National Social Science Survey as analysed by the Christian Research Association.
5. Blombery, 'Tricia & Hughes, Philip J., *Patterns of Faith in Australian Churches*, Christian Research Association, Hawthorn, 1990, p. 20.
6. In *The Catholic Worker*, June 1986, p. 12.

that people have become far too materialistic. [7] Yet, while Australians have seemingly disowned the churches, there has been stability in the numbers of people who believe in God and in spiritual values. In fact, belief in the transcendent has been so constant that sociologists Garry Bouma and Beverly Dixon argue that there has not been any significant loss of religious influence in Australia and that Australia is basically a religious country. [8]

Catholic reactionaries blame the drop in practice rates squarely on the 'confusion' created by the 'liberal' clergy and laity and the failure of educators to inculcate 'correct' Catholic doctrine and discipline, especially in church schools, seminaries and religious orders. More remotely, some reactionaries blame the consequences of the Second Vatican Council. They claim that if only Catholics returned to 'right' doctrine, restored the past and obeyed the pope and bishops (but especially the pope), a new 'golden age' of Catholicism would return. Vocations would increase, Mass attendance would return to the halcyon days of the 1950s and the church would once again stand over and against a world falling apart.

Bishop George Pell, the auxiliary bishop of Melbourne, espouses this view. He says that he belongs to the 'Irish school of romantic conservatism' — whatever that might be. He has recently argued that Catholicism will

slow the exodus from the Church and attract more converts not only by defending and developing the core of Catholicism, but also by energetically promoting many aspects [of belief] . . . which we have been tempted to downplay for reasons of ecumenism or as concessions to modernity . . . I urge a style which is a mite more confrontational and certainly much less conciliatory toward secular values. The Cross is a sign of contradiction. [9]

That sounds reasonable enough, but Pell then goes on to outline the type of traditional Catholicism that he thinks should be revived. Despite favourable comments about changes in the church over the last twenty-five years, the picture he sketches seems little different

---

7. For an interesting study of the question of secularisation of religion in Australia see Ireland, Rowan, *The Challenge of Secularisation*, Collins Dove, Blackburn, 1988.
8. Bouma, Garry D. & Dixon, Beverly R., *The Religious Factor in Australian Life*, MARC, Melbourne, p. viii.
9. Pell, George, *Conversazione: Catholicism in Australia*, Seminar on the Sociology of Culture, La Trobe University, 1988, p. 10.

from pre-Vatican II Australian Catholicism in the 1950s. What the bishop is actually calling for is the restoration of a Catholic sub-culture that stands over and against the wider world.

Australian Sociologist Rowan Ireland has recently pointed out that responses to secularisation like those articulated by Pell, are becoming increasingly common in the Catholic church. This is how Ireland characterises the anti-secularisation response of the reactionaries:

[The] response also includes centralisation of authority and an attempt to check post-Vatican II disarray through unity around the traditional hierarchy. Specialised apostolates, especially those forming militant elites among the middle and upper classes, are restored. The cognitive dimension of religion, and within that the propositional mode, is construed, despite the protests of educators and pastors in the field, as the springboard to renewal. [10]

Ireland then makes the very penetrating observation that the retreat to the sub-culture leads directly to the church aping the very secularist agencies that it so vigorously attacks.

The church becomes another specialised bureaucracy, manned by experts. In organisation and alliance it comes to bless and embody materialist assumptions, despite disavowals of rhetoric and ritual . . . the church loses its autonomy as it becomes a look-alike of and a socialising agent for dominant economic and political institutions . . . The church becomes its own gravedigger. [11]

Ireland, together with Paul Rule, develops these ideas further in an essay in Alan Black's *Religion in Australia*. [12] Firstly, it is clear from Ireland and Rule's evidence that Australian Catholicism is not — and probably never was — a monochrome monolith. Catholics simply do not accept the official position of the hierarchy on a whole range of issues, such as contraception, divorce, homosexuality, euthanasia and even tax cheating; in all surveys on these topics, Catholics are close to community norms.

Ireland and Rule discern two responses of Australian Catholicism to what is happening in the wider world:

---

10. Ireland, op. cit., p. 45.
11. Ireland, op cit., pp. 45–46.
12. Ireland, Rowan & Rule, Paul, 'The social construction of the Catholic Church in Australia' in Black, Alan W., *Religion in Australia: A Sociological Perspective*, Allen and Unwin, Sydney, 1991, pp. 20–38.

On the one hand there are important individual bishops and groups whose major project is that of 'recentrage' — the restoration of an hierarchical line of command and the centralisation of decision-making in a Church which, according to the apostles of recentrage, has fallen into disarray since Vatican II. On the other hand, there are experiments in decentring: the transformation of the Church into a flatter, more diffuse structure, a loose network of communities in which a rich variety of ministries is recognised and the boundary between clergy and laity less marked than at present. [13]

Ireland and Rule also show that even within these two general approaches, there are many variations. Among the recentrage projects are Opus Dei, the AD 2000 group, the John XXIII Fellowship, the Newman Graduates in Sydney, and single issue groups such as Right to Life and CARE (which is concerned with Religious Education). Bishop Pell has emerged as a major spokesman for the recentrage project. He has said clearly that the church must return to a parish-centred ministry with large Mass attendance and traditional devotions, all under the strong leadership of the parish priest. [14]

In my view the analysis of the reactionaries is superficial. The factors that they delineate are, at most, only minor causal issues in the decline of the influence of the church. Their reactionary solutions to the problems of the church simply will not work; you cannot retreat to and restore the past.

The real reason for the crisis facing the church lies deep within the life of the Australian Catholic community itself. It begins with the fact that the church has no clear self-definition and that, as a result, it cannot articulate what its role is in Australian society. It simply has no idea where it is going. When I say the church has no 'self-definition', I mean that it has lost its way in this society, that it has no sense of its function and purpose. Uncertain of their identity in Australian society, it is natural for some people in the church to retreat into a sub-culture where they feel safe, within the narrow confines of which they can define themselves and assert their identity. The problem is that this narrow definition excludes the majority of the best and most creative followers of the Catholic tradition.

The key people in the articulation of goals in the context of a church that is still fundamentally hierarchical, are the bishops and the clergy. But there is little or no leadership coming from the

---

13. ibid., p. 30.
14. Pell, George, 'The Future of the Local Parish', *AD 2000*, 2/9(1989), pp. 10–11.

hierarchy. In fact, as Pell's comments show, there is a tendency among some of the articulate bishops to constrict the boundaries and to turn the church inward, away from participation in wider society. Much of the clerical leadership and some lay members of the Catholic community seemingly want to retire into a sub-culture where they feel safe and cossetted from the hard questions. Rather than seeing the church as a leaven within our society, they prefer to define it tightly and have it stand over and against Australian culture.

Certainly since Vatican II, the laity have been called upon to take a more active role in the life of the church, and some have responded positively. But these people are still only a small minority and, unless they have influential positions in Catholic educational or diocesan structures, they are powerless. In fact, as Kevin's story illustrates, it is often those who have tried hardest who feel most marginalised.

On the national level in the Australian church the real decisions are taken by a small coterie of bishops and their immediate advisers. Power, and the even more important sense of participation in decision-making, is a little more widespread on the diocesan level in some places, but the structures are still very centralised. The large majority of Australian Catholics have a strong personal sense of 'being' the church and belonging to the Catholic community. But, at the same time, because they experience alienation and marginalisation, they know that they do not participate in the real decision-making processes. A recent example of this is the establishment of the Australian Catholic University; there was no broad consultation about the establishment of this institution. I will return to the story of its foundation in more detail later.

The theme of sub-culture will recur in this book. By this I do not mean that the church has retired to the historic Australian Catholic 'ghetto', which, in any case, was often more 'Irish' than Catholic. The simple fact is that since the Australian Catholic church is now truly multi-cultural in its composition, there is no ethnic ghetto to which it can retire. Also, when I say that the church is a sub-culture, I do not mean that it has become a sect, in the sense of a small, closed, inward looking group that sees itself as deviant, with a viewpoint that is over-and-against the wider society.

Rather the Australian Catholic church has become a sub-culture in the sense that it has retired within its own borders, turned back on itself, withdrawn from any serious engagement with Australian society and unconsciously assumed that it is sufficient to itself. The questions that it addresses and the language that it uses are not those of contemporary Australia. Australian culture is passing it by, for it is perceived by many to have little to offer, except romanticised nostalgia for the memories of a Catholic boyhood or girlhood. Fundamentally, the church's failure is in the task of translating the message of Christ and the Catholic tradition into this culture.

I am not saying that all Catholics are sub-cultural, or even that a majority are. Rather, the problem lies primarily with many in the church's leadership. They are simply out of touch with where most Australian Catholics are in their religious and human development. People are not going to Mass because they experience there little that engages them or makes sense to them and to their experience. Many of those who have ceased going to Mass are not materialistic, or selfish, or secularised. Most are quite serious about human and spiritual growth and they look hopefully to the church for support in this. But they find nothing except bad liturgy and a superficial moralism that sometimes even contradicts the true Catholic tradition, and they hear the questions that they think are important treated with little less than contempt by church spokesmen. This is especially true of Catholic women. It is significant that the decline in the practice of women has been much greater than that of men.

Sexism and clericalism convey to women the reality of exclusion. This came home to me strongly in the period after the publication of *Mixed Blessings*. I had mentioned the issue of sexism in the book, but one reader accurately commented to me that I conveyed no sense of pain or the passion of women. She told me in a letter:

You convey no sense at all of the extraordinary rage, frustration and sheer pain that so many women are feeling. Don't you know what it feels like to be at Mass, sunlit and splendid, and know yourself at one with a God who is in pain as the endless tide of male self-glorification rolls on — despised or invisible in the language of the Old Testament, the New Testament, the homily . . . and even in the words of the liturgy itself . . . [subjected to] images of omnipotent maleness.

Another example of the passion and pain is provided by Kathleen Curmi. She is 57 and describes herself in the book *Sweet Mothers, Sweet Maids* as a 'practising Catholic'. But she also says that 'I have become increasingly aware and angered by the sexist bias against women in the church'.[15] As an illustration, she describes a time

---

15. See her story in Nelson, Kate & Nelson, Dominica, *Sweet Mothers, Sweet Maids: Journeys from Catholic Girlhoods*, Penguin Books, Ringwood 1986, pp. 125–131.

when she and her husband needed some help with the temperature method of family planning. They went to the local Catholic Family Welfare Bureau:

We were interviewed by the expert. What a priest! His arrogance and sexism were unbelievable. He addressed all his questions and remarks to my husband. It was as though I was an efficient incubator but had no power of speech, let alone a brain. It was the one occasion in my life when I felt really alienated from the church.

Many women of passion and conviction feel that there is no future for them in the Catholic Church. It is a tragedy that the reality of the church forces them to experience this.

The simple fact is that the Australian Catholic church, the largest religious community in this country, has lost its way. And by losing its way, it may well lose its opportunity to participate in shaping the crucial decade leading up to the coming millenium.

## *Kairos* — the time of decision

At the same time the Australian church today is confronted by a marvellous opportunity. In the 1970s and early 1980s there was a feeling abroad in North America and Europe that a crucial time for Catholicism had come. A moral and religious vacuum had been left by the decline of mainline Protestantism and secular liberalism. The Catholic church had been through the difficult and trying period of the Second Vatican Council and had seemed to have emerged with its tradition intact. But it also had sufficient openness to have something substantial to offer to the contemporary western world. This was articulated recently by Peter Steinfels in *The New York Times*. [16]

Possessed of a sophisticated style of moral reasoning that is rooted in religion but philosophical enough to engage non-Catholics and perhaps even agnostics and atheists, the Catholic Church . . . seemed poised to make a unique contribution to the country's public life.

In Australia this sense of opportunity has never been articulated as clearly as it was in Europe and North America, but certainly the

---

16. Steinfels, Peter, 'State of the Church: Has Catholicism Lost a Chance to be Our Moral Clearinghouse?' *The New York Times*, Sunday 8 April 1990.

chance is there for the Australian Catholic church to participate in public dialogue about social policy, and to play an active role in the formation of new cultural and societal attitudes.

This opportunity is especially pressing in the area of public morality and social justice. The decline of traditional values, largely inherited in the English-speaking world from Protestant Christianity, had created a vacuum in public moral thinking. Australian secularism, which had provided much of the ideological underpinning and national myth for most of the 20th century, has degenerated into crass free market materialism. This has occurred just at the time when Catholicism has emerged as the largest single religious community in Australia. The Catholic church has a sophisticated moral tradition that has been further developed in the contemporary world by a group of excellent moralists, many of whom were born and bred in English-speaking countries (especially in the United States). The Australian Catholic community has the potentiality to contribute much to public discussion of issues such as war and peace, medical ethics, social justice, public and business administration and, most recently, environmental ethics.

The bible, of course, has a marvellous word for this moment of opportunity. It calls it the *kairos*, the time of decision. The word also refers to the moment of crisis when a wrong decision can be taken, but at heart the word points to the creative crisis of change. It refers to the special moment when hard decisions have to be taken. This is highlighted even more by our entry into the last decade of the 20th century and by the coming millenium.

In some parts of the world, the church has, of course, seized its *kairos*, its moment. In Latin America Catholicism has become an agent of social and religious change. The theology of liberation has provided the superstructure, but it is at the grassroots level of community attitudes that the real change has occurred. In the process the Catholic church has become one of the truly creative forces in Latin American culture. A similar phenomenon has happened in sub-Saharan Africa where the Christian churches generally, and the Catholic church particularly, are providing African culture with the linguistic, philosophical and theological context that it needs to come to a fuller development.

The *kairos* has also come for the Australian church. Catholics face the creative crisis of opportunity. Decisions need to be taken that will determine the future of the church in this country. Contemporary Australia is a questioning society, seeking identity, asking questions of meaning, looking for spiritual answers. The bankruptcy of individualistic, free market materialism provides an opportunity for an ethic based on social justice. It is a unique time if only Catholics can discern the real questions and examine them in the light of the rich Catholic tradition of theological and moral reflection.

But, tragically, leading Catholics seem to be retiring into a safe and unquestioning sub-culture. The church is perceived by many Australians as socially and morally unsophisticated. It seems to them to focus on a small number of issues, usually connected with sexuality and generally in the most restrictive way. Its reasoned and profoundly Christian moral, social and ethical insights are unknown, except by a tiny coterie of experts and the better educated clergy. It is significant that Peter Steinfels maintains that the same phenomenon is happening in the United States:

Today talk of the Catholic moment is fading; and with it, the hope that the church could bring its history of reasoned discussion to a range of baffling issues that include nuclear deterrence, medical ethics and biotechnology . . . The leadership of the American church has grown shy of innovation. While some maintain that this newly imposed doctrinal discipline is the very soul of sound churchmanship, to others it is a recipe for calamity.

The tendency to retire into self-enclosure is further reinforced by the agenda of Pope John Paul II, who seems to want to restore the church to the status of a 'perfect society', a concept much promoted by the popes of the 19th century, especially Pope Pius IX. The idea of the church as a perfect society goes back to the Middle Ages. For John Paul II the concept means that the church is an institution that is complete in itself and to which the state must bow in religious matters. In this view, the church sees itself as the sole source of spiritual and cultural salvation for all. This, of course, has grave implications for the idea of religious freedom. The irony is that for John Paul 'religious freedom' means freedom for the Catholic church to operate without external restraint.

The revival of such out-dated ideas distracts the church community from its real task, which is to be a leaven in the contemporary world. It reinforces an emerging Catholic fundamentalism and forces the church back into a sub-cultural self-definition. As a result, Catholicism is inclined to turn away from the wider world which, right now, has an acute need for a religion and a spirituality which provides a sense of meaning and a rational, humane morality.

It is not only in morality that the Catholic church has much to contribute. The Religious Department of the ABC is in close contact with a wide range of religious thought and activity across Australia. Our experience is that in Australia — as in other western societies — there is a profound 'spiritual desire' that affects many people. They look for a deeper experience of themselves and others and a more coherent sense of meaning for their lives. But this spiritual desire is not expressed in traditional language or through the kind of questions to which the church has ready-made answers. It can only be perceived obliquely, often in terms of feeling better about oneself or in terms of wholistic well-being or through an attraction to non-Christian or non-traditional spirituality. Certainly, much of this is self-engrossed and narcissistic, but it is often waiting to be informed by a reality with more depth that will give it a more profound shape and direction. Even in the crasser manifestations of new age religion you can discern a primitive spiritual longing. In its altruistic forms this longing expresses itself in a desire for social justice, for the care of our dying planet and all its living species, as well as in the more self-involved longing for personal fulfillment.

But the representatives of the Catholic church in Australia continue to talk a language that is fundamentally arcane and foreign to contemporary Australians and to the needs and questions that they express. Most senior Catholic churchmen project an image in the media of people from the past, whose verities are irrelevant to present-day life. They seem to have little perception of the profound spiritual questioning that is moving through our culture. Paradoxically, that supreme churchman, John Paul II, who is such a master of television imagery, speaks a rhetoric that sounds modern, but in reality he is addressing the questions and preoccupations of other cultures and ages past. [17]

I believe that the 'Catholic moment' is rapidly passing in Australia. The sense of direction and theological and spiritual vigour that the Catholic tradition has to offer has been effectively stymied by the very ones who believe in it most, but who are too fearful to confront the contemporary questions that demand that they move outside their known Catholic sub-culture.

I now want to examine this 'Catholic moment' in the light of what has happened to the major groups that make up the Catholic community: bishops, priests, religious and laity.

---

17. I have analysed the dangers inherent in Pope John Paul's media style in 'The Peripatetic Pope' in Küng, H. & Swidler, L., *The Church in Anguish: Has the Vatican Betrayed Vatican II?*, Harper & Row, San Francisco, 1987, pp.52–57.

❖ *2* ❖

# *THE PEOPLE*

## *The bishops*

I wrote this section of the book during the first week of the western and Arab response to the Iraqi invasion of Kuwait. Many thoughtful Australians pondered deeply the implications of our nation's participation in this war. Even in the much-criticised media, serious moral questions were asked about the justification of declaring war and about the actual conduct of the campaign. The 'just war' theory was invoked by both sides of the debate. Peace groups formed and politicians made a show of debating. But the bishops of the largest church in the country, with a long and sophisticated moral tradition, had nothing official to say as a group. Certainly individual bishops spoke, especially after the war began. But the vital time was before the beginning of hostilities. I was present as the World Council of Churches Assembly in Canberra agonised over the morality and implications of the war, but after initial tentative statements, the Australian Catholic bishops fell silent. Not a murmur was heard from them.

This was in sharp contrast to Pope John Paul II who, in an Angelus address on the day after the war began, said: 'Never have the leaders of nations, servants of the common good, been so challenged by their consciences'. The pope's moral judgement on the war was unequivocal:

The use of military force to dislodge Iraqi troops is not justified, because of the massive loss of life and destruction that it would produce. Diplomacy and non-violent means are the way to deal with the situation.

How prophetic the pope's words have proved to be! If the application of the principles of the just war was too complicated, then perhaps the bishops could simply have quoted Jesus: 'You have heard that it was said, "You shall love your neighbour and hate your enemy". But I say to you, Love your enemies and pray for those who persecute you' (Matthew 5:43–44).

Admittedly, it all happened in early January, when Australians are traditionally on holidays! I actually tried to contact ten separate moral theologians, both Catholic and Protestant, seeking an on-air comment about the morality of war, only to be told that all ten were 'on holidays'! So the bishops were not alone in their tardiness. Certainly, Christian pacifism and moral idealism were well represented by the prophetic action of the few who simply went and stood in the firing line. It was also well represented by the reasoned and strong moral statements of churchmen like Archbishop Keith Rayner of Melbourne, then Acting Primate of the Anglican Church, and by the leaders of the Uniting Church. But what did we hear from the Catholic bishops? As a group, nothing; as individuals or through spokesmen, a few words about 'just war', then silence.

This war, like a number of other major issues being discussed in Australia at the present moment, raised ethical, social, economic, political and environmental issues. With a long tradition of social morality you would expect that the Catholic church and its forty-three bishops would have something to say. You would expect that they could provide some basic moral principles upon which discussion could be based and that Catholic experts on the morality of war would be encouraged to participate in public debate. Sure, some of the bishops attended the peace services and got their pictures on television and in the newspapers. But they said little and were not prominent in any public debate.

## Engaging in public discourse

It is true that, at times, the bishops, both individually and as a group, have published statements on various topics. As was pointed out to me by Monsignor Kevin Manning, then secretary of the Bishop's conference and now the Bishop of Armidale, when I made a related accusation in a recent article, the bishops have issued a number of statements on issues of contemporary importance in Australia. [18] Later I will praise the bishops' draft statement on wealth and poverty in Australia, *Common Wealth and Common Good*, especially for the process that they used in developing it. [19] However, this is the exception rather than the rule.

In my view the problem is at a deeper level than that of the issuing of statements. It lies in the Australian Catholic bishops' apparent inability to enter into what might be called 'public discourse'. Democratic societies have evolved fairly sophisticated skills in the formulation of public policy, as well as an ability to balance the conflicting and often contradictory expectations of the many competing groups that make up complex modern, elective, parliamentary states. This is achieved through public discourse.

Genuine leadership in our society does not just involve issuing statements. It means participation in a broad public dialogue about those statements. Of course it involves having convictions and being willing to express them. These convictions can be passionately held and strongly expressed. But public dialogue demands that participants not only hear other views, but respect the right of others to hold their views, however vigorous the disagreement. It implies a willingness to play the democratic game and to participate in the political process. However, this creates considerable problems for a church that makes claims to possess 'Absolute Truth'.

Underlying our democratic process is a genuine sense of the 'Public' (with a capital 'P'). By the Public I mean the wide diversity of views which make up the total community of which the varying pressure groups are component parts. [20] In democratic societies the people are informed of the range of views that compete in public discourse through the media. The media is the key mediating element in the process of debate from which public policy and opinion can emerge.

Thus when I say that the Australian church lacks leadership, I am referring not only to the failure of the bishops to address a key issue like the participation of our nation in war, but also to the

---

18. See 'In Search of the Church: Catholics at the Crossroads', *National Outlook*, July 1990, pp. 6–11. Monsignor Manning's letter is found in *National Outlook*, September, 1990.
19. *Common Wealth and Common Good: A Statement on Wealth Distribution from the Catholic Bishops of Australia*, Collins Dove, Blackburn, 1991.
20. For a good treatment of these issues see Parker J. Palmer, *The Company of Strangers: Christians and the Renewal of America's Public Life*, Crossroad, New York, 1983.

weakness of Catholic leadership in engaging with other views in serious media debate. Church people constantly complain that they are treated badly by the media, that there is a secularist-liberal bias in Australia in television, radio and print. There is some truth in this accusation, although I think that journalistic ignorance about religion, and bad performances in the media by some Catholic leaders, are the real problems. There are very few people in the media who know anything about religion, although one cannot help noticing an extraordinary number of lapsed Catholics in the industry. Outside the ABC's religious department, I can only think of a handful of journalists in general media in the whole country who are professionals in religion, let alone experts in Catholicism.

Allowing that the media is far from faultless, the church also needs to be honest about the significant failure of Catholic leadership in coming to grips with the need to present a more humane, accessible, Christian face. Often the church is represented by nervous, unsmiling, apparently inhumane men who project an image of moralistic simplicity in the face of complex human and ethical issues.

Let us take an example of this from late 1990: the debate about federal government regulations affecting Catholic nursing and retirement homes. The Bishop's Conference claimed that, by allowing residents the right to bring in their own doctor, the government was opening the way to medicos who did not agree with the Catholic teaching on euthanasia. (At least this was an advance from the initial stage of the debate which focused on the sexual activities of the aged residents.) I am not questioning that direct killing is wrong — governments recognise this as well, since there is law against murder! But the Catholic moral tradition has always recognised the complexity of questions concerning the right to die and has sophisticated norms for the use of extraordinary means for maintaining life. But nothing of this wealth of Catholic tradition was articulated by Catholic spokespeople. The discussion was reduced to blank assertions that 'Thou shalt not kill'. The Catholic position not only appeared unsophisticated, but, much worse, it appeared unsympathetic and lacking in compassion. If any issue cried out for delicacy and skill, this one did. None of it was in evidence, and the media was not to blame for that. The worst casualty was the Catholic moral tradition itself.

The Catholic church's leaders are good at issuing statements and even pontificating upon them, but they are generally inept at debating and defending their position in the face of conflicting opinions and attitudes. And they often show scant respect for conflicting views. Much of this attitude originates in the church's unspoken but profound ambivalence about living in a liberal democratic society. It is hard for those trained in hierarchical ways of thinking, who are used to the exercise of ecclesiastical power, to engage in discussion with those who assert that their views are of equal weight and importance to those of the institutional church. It is because they have avoided this public discourse, or have been unwilling to present, or even worse been ignorant of, the whole Catholic tradition, that I criticise the bishops and church leadership for having seemingly abrogated their role of speaking for Catholicism.

There has been one significant and very important exception to all this. The bishop's draft statement on wealth distribution in Australia, *Common Wealth and Common Good*, is a fine document that takes the church right back to centre stage in the debate about important moral and social issues in this country. And the process of the evolution of the document is as important as the document itself. It is the product of a three year process of consultation, with public hearings in major cities and the sifting of over seven hundred written submissions. There was a genuine attempt by the bishops to involve the Catholic community in the formulation of the document and thus in the development of Catholic social teaching.

The great strength of the document is its underlying premise: it re-asserts the Catholic principle of the corporate nature of the human community. Society is not made up of a mob of selfish individuals, struggling to drain as much money as possible from the system — and usually contributing nothing in return by minimalising taxes. The bishops see society as a corporate reality in which all human beings have a right to share in the goods of creation. The evidence of a loss of a sense of the mediating role of society was revealed in a speech of former British Prime Minister, Margaret Thatcher, to the General Assembly of the Scottish Kirk some years ago. She told the assembled divines that there was no such thing as 'society'; just individuals, families and the government! This, of course, is founded on the doctrine of economic rationalism, which argues that wealth is created by maximising the freedom of individuals to exploit the economic opportunities available. It seems more like the law of the jungle than a form of rationalism. It has no real concept of the corporate nature of society and is little more than a justification for selfish individualism. Such a view is totally contrary to Christian teaching and, in my view, cannot be morally justified by the Catholic tradition of social ethics.

Australian society, especially since the 1980s, has been bedevilled by great inequality in the distribution of wealth. Our governments have also encouraged so-called 'entrepreneurs', most of whom have now proved themselves to be little less than common thieves. By restating the Catholic tradition on the corporate nature of society and the interdependence of all individuals within it, and by stressing the mutual obligations that people have to each other, *Common Wealth and Common Good* is the best contribution that the Catholic bishops have made to the discussion of social ethics in Australia for a long time.

On some other public issues, the bishops have made stands. Certainly, one thing the entire episcopate has been consistent on is the issue of abortion. Generally speaking, too, the bishops have supported Aborigines and their claims for a fairer deal in Australian society (although, in practice, some senior archbishops quickly modified their commitment to the Aboriginal cause when it became clear that this would involve them in confrontations with certain premiers and state governments). They have supported Aborigines with strong statements. The Aboriginal issue has been pushed along by both public opinion and a determined lobby of clergy and laity. The episcopal spokesman on this area, Jesuit Father Frank Brennan, keeps the issue before both politicians and the media. Medico-moral issues have also been discussed publically by the Church, especially in Melbourne where the in vitro fertilisation programme originated. This discussion has occurred because there is an effective and thoughtful lobby of Catholics publically discussing the issue, such as Salesian Father Norman Ford (*When Did I Begin?* (1989)), Dominican Father Anthony Fisher (*IVF: The Critical Issues* (1989)), and medical ethicist, Nicholas Tonti-Filippini, now moved to Canberra as a research adviser to the Australian bishops.

The area of research is a problematic one. The only truly independent research organisations within the Catholic church are run by religious orders. Through Tonti-Filippini, the bishops have commissioned a number of research projects. Recently, there has been some consultation as to what issues ought to be tackled, but the focus will probably continue to be on medico-moral questions. However, the Catholic church does support the ecumenical Christian Research Association, which has already produced good material on the beliefs and values of Australian Christians and of their attitude toward the clergy. A number of metropolitan dioceses have their own pastoral research offices, such as Melbourne, Brisbane, Adelaide and Perth. While these may help to shape policy decisions in the individual dioceses, they have no overall role to formulate policy for the national church. Sadly, in some cases it is even doubtful if local dioceses take a great deal of notice of them!

This lack of leadership on the national level is also shown in intramural church issues. There has been no serious research into the actual ministry of the Australian church. As a result, there is no plan or range of priorities for the future direction of Catholicism. Major changes are occurring within the church that affect the ministry seriously: the Australian clergy is ageing, religious orders are not attracting new members, the laity are becoming more involved in intramural church ministry. The Melbourne Catholic Research Office for Pastoral Planning has done some research on lay pastoral associates, but outside of this there is little reflection on the consequences of these shifts and there are seemingly no plans to deal with the issues that result. One hopeful sign is that courses are beginning to be offered to train people for particular ministries. The Catholic school system continues to exist as a major component of the ministry of the church without any articulated philosophy about its purpose and with little research concerning its effectiveness in forming the faith of its students. [21] The failure to confront these issues is a failure in leadership.

I certainly do not claim that the bishops and their secretariat should have to do all of these things themselves. It seems to me that Christian leadership demands that they support and facilitate qualified and committed people to do the work. The task of Christian leadership is to stimulate and give focus to the gifts of others. Yet the opposite seems to happen. Those who try to articulate the major issues facing the church are often seen as a threat and discouraged — or even accused of 'hating the church'!

## Pastoral successes

It would be unjust to leave this topic without mentioning those bishops who have really tried to involve laity, clergy and religious in the process of planning for the future. The archdioceses of Brisbane

---

21. An exception to this is Marcellin Flynn's *Catholic Schools and the Communication of Faith*, St Paul Publications, Homebush, 1979. I will refer to this later in the section on Catholic schools.

and Adelaide, for instance, have shown considerable sophistication in the training and integration of sisters and laity in a range of ministries. Both archdioceses have also provided ongoing education for their clergy. The Archdiocese of Canberra-Goulburn held a diocesan synod in 1989 which involved a cross section of laity, clergy and religious from all parishes in the archdiocese which extends from Lake Cargelligo in the central west of New South Wales to Pambula on the south east coast and Bombala, on the edge of the Snowy Mountains. The Synod set out a pastoral plan that will take the church in Canberra-Goulburn into the next century.

Several bishops, such as Bishop Bede Heather of Parramatta, have excellent reputations for their pastoral initiatives and ecumenism. In Queensland most of the bishops are noted for genuine pastoral creativity. Archbishop Leonard Faulkner of Adelaide has achieved something quite unique in Australia: he has set up a Diocesan Pastoral Team, a small group comprised of laity, sisters and clergy that acts as his immediate advisory body and which assists him in his episcopal ministry and attempts to 'get things done' in the archdiocese. In this way — whether intentionally or not — he has side-stepped the clerical monopoly on diocesan power. In Adelaide there has been discussion of the effectiveness of the team and even a recent assessment by an outside bishop (Bishop Ronald Mulkearns of Ballarat) and a priest (also from Ballarat). There are not many bishops anywhere in the world who would allow one of their fellows to assess critically the performance of one of their key diocesan structures!

The great strength of the Australian episcopate is that a number of the bishops are very pastoral in their attitudes. In fact, most of them seem to act very much like parish priests. They apparently conceive of their ministry as directed primarily to the pastoral nurture of those in need. By 'pastoral' I mean that they genuinely care for their people and communities and they are willing to work hard at the ministerial level. In other words, on the micro, local level, the majority of the Australian episcopate are good pastors.

It is true that there are extreme situations which will bring out leadership potential in persons who may not have shown it. In situations of persecution, church leaders have discovered inner strength and power, such as Archbishop Oscar Romero of San Salvador, or Bishop Carlos Bello in East Timor. Australia, thankfully, is not haunted by death squads nor ground down by military occupation. However, we still have an obligation to look for genuine leaders among our bishops!

But we will not get genuine leadership if the present dispensation for episcopal appointment in Australia continues. The method of election ensures that strong, out-spoken people are excluded. Those with the best chance of joining the episcopal bench are quiet, hard-working, pastoral and institutionally loyal priests. There have been occasional exceptions to this, but the general pattern is of men who have not spoken out on major issues. This means that the group best suited psychologically to provide strong, positive leadership is usually passed over. The present method of election certainly helps to maintain Vatican control of the local church, but it does nothing for genuine episcopal leadership. The pivotal person in episcopal appointments is the Apostolic Nuncio, or papal ambassador in Canberra. He is almost always a career diplomat. With the exception of a former chaplain to the Belgian royal family (Archbishop, later Cardinal de Furstenberg), all have been Italians. Historically, the Australian bishops have often shown ambivalence toward papal representatives. Archbishop Daniel Mannix of Melbourne and Archbishop James Duhig of Brisbane believed firmly that they should be kept in their place. Historian Father Tom Boland stresses that both Mannix and Duhig tried to maintain a balance between the power of Rome and the authority of the local church. [22]

The present Apostolic Nuncio, Archbishop Franco Brambilla, has caused considerable annoyance to the local episcopate, as well as to the leadership of the religious orders of both men and women, through his ill-informed and undiplomatic comments and his attempted interference in a number of internal matters. It is asserted that his major sources of information about the Australian church are right wing reactionaries and that the nunciature has provided a channel to the Vatican for the complaints of these people, who in no way represent the general Catholic community. The Nuncio's task is to see that Roman policy is carried out. This policy insists on a strict adherence to the priorities of the present pope, especially in debated issues such as the celibacy of the clergy, opposition to contraception and abortion, loyalty to the full range of non-infallible papal teaching and at least verbal commitment to social justice. However, despite the fact that it is the Nuncio who draws up the 'terna' or list of three names of episcopal candidates for vacant

---

22. Boland T.P., *James Duhig*, University of Queensland Press, St Lucia, 1986. See especially pp. 210–212.

dioceses or for possible appointments as auxiliary bishops, the influence of the senior members of the Australian episcopate is still very considerable, and, in most cases, is probably definitive. These bishops are essentially cautious and conservative. Their personalities and attitudes have been formed by years of institutional involvement. To a considerable extent they reflect the clergy from whose ranks they were drawn. Like all normal human beings, bishops are wont to choose people like themselves as their colleagues and successors. So the system is geared to exclude strong, creative personalities. Probably there is something very Australian in it all: we are essentially suspicious of strong individual leadership of any sort.

It is worth noting in conclusion that present Roman policy, as articulated by Cardinal Josef Ratzinger, is to deal only with local bishops as individuals and to deny any genuine ecclesiological status to national Bishops' Conferences. This is a clear attack on the theology of Vatican Council II, which gave expression to the traditional Catholic teaching that the bishops govern the universal church in solidarity with the Bishop of Rome. It is what the council called episcopal collegiality. Despite its theological camouflage, this Roman policy is obviously rooted in the old pagan dictum *Divide et impera* — divide and rule. However, Cardinal Ratzinger need not worry too much about the Australian Bishop's Conference, as he apparently does about episcopal conferences in places like Brazil and the United States. At the present moment it seems unlikely that the bishops of Australia will ever do anything too radical, let alone confront Rome on any important issue!

# *The clergy*

If you want to be a practising Catholic, at least practising in the eucharistic and sacramental sense, you will have to deal with the clergy. They are the most important leadership group in the hierarchical church because of their direct impact on people at the local level through parishes and religious communities. In Australia there are 2040 diocesan priests and 1301 priests who are members of religious orders. [23] (Although, as I indicated in *Mixed Blessings*, the statistics in the *Catholic Directory* are somewhat inflated by counting priests who are retired, sick or on leave).

---

23. *Australian Catholic Directory 1990–1991*, E.J. Dwyer, Newtown, 1990.

There is considerable agreement among overseas commentators that the Catholic clergy is a profession afflicted by an identity crisis. This is a lot more serious than the malaise which affects many of the contemporary professions. The role of the clergy has changed radically since the Second Vatican Council. Invested with a share in divine and ecclesiastical power through ordination, their role before the Council was 'to shepherd the flock', and in Australia some priests had the reputation of doing so in no uncertain terms! However, over the last twenty-five years, priests have been asked to give away many of the ministerial tasks which provide meaning for their lives and to take on a non-directive leadership role for which none of them were trained and, as we shall see, very few are psychologically prepared.

I shall never forget the bitterness of one older priest with whom I worked in a parish. He deeply resented the incursions of the laity into what he regarded as the preserve of the clergy ( and he meant such ordinary things as taking communion to the sick or helping with marriage preparation). He often said that if he had thought that married lay people could do what he had been ordained to do back in the 1930s, he would have never entered the seminary. His whole personal identity was tied up with his role as a priest. It was clear to him that his task was to assist the bishop 'in ruling and shepherding the flock entrusted by God'. I think the level of pain among many clergy as they face the changes occurring in the church is often underestimated. There are certainly many grounds for criticising priests, but the Catholic community should also be aware of what is being expected of them: to surrender major aspects of the only job that some of them know. That is asking a lot of ordinary human beings!

Right from the start, I need to make a confession: I am a failed church leader myself. For two and a half years I was a parish priest. This is not much of a record, given that I have been ordained for twenty-four years. I soon got out of leadership and returned to the much easier job of lecturing others on how to do it! But during my period as a PP, in the large and marvellous Sydney parish of Randwick, I learned one thing: it is much easier to be a dictator than a participative leader, and it is much simpler to do everything yourself than to involve others. Despite complaints about paternalism, most people respond better to benevolent 'guidance' than they do to the challange to participate.

But once people do begin participating, a genuine leader has to take them seriously, to listen to and value their opinion, even when there is a conflict with the leader's own priorities. This is exactly what happened to me at Randwick. What the parish committees thought was important and what I thought were priorities were sometimes two very different things. Often also the priest has to reconcile and give direction to all of the differing points of view in the parish community. So while I criticise other priests, I have not forgotten my own brief stint in Christian leadership.

## The problems facing priests

I have argued that there is a direct connection between the Catholic Church's retreat to the sub-culture and the failure of its leadership cadre. There is a peculiar twist to this in Australia: we are inherently suspicious of anyone who has the talent of leadership. The myth of egalitarianism is a powerful force in the national psyche. Australians love to 'bring down the mighty from their thrones' (unfortunately we are not so keen on 'exalting the lowly'). This is seemingly all very democratic, but the problem is that we often lack the discernment to distinguish between the power-hungry self-seeker and the genuine leader. Also Australia has had such a mediocre tradition of leadership in both state and church, that we probably could not tell a real leader if we met one! Australian Catholics are so tuned to a hierarchical mode that they constantly expect the bishops and clergy to be the only persons in the church who could possibly be leaders. Tragically, that is probably the last place to look.

A major leadership problem is that so many talented men have left the ministry. Catholics should underestimate neither the number nor the calibre of those men who have left the priesthood. The majority are very successful in their new careers in business, social service, teaching, media or other areas. A significant proportion still work in areas that are as ministerial as anything that they would do in a parish. Part of the church's leadership crisis can be traced back to the high attrition rates among the clergy, an attrition that is due largely to unfulfilled celibacy and to a lack of vocational satisfaction. The fact is that the Australian church has lost a whole group of men who would be natural leaders and the dearth of leadership in the church now reflects this.

Certainly there are still creative men left among the Australian clergy who are trying to minister realistically in the contemporary world. But their creativity is often stifled by the demands and expectations of the institutional church and occasionally by the vicious criticism and opposition of some reactionary Catholics. This has been experienced especially by priests who have stepped out to minister to marginalised groups, such as Aborigines, or who have encouraged theological, catechetical or structural change.

In fact, there seems to be a growing anti-clericalism among reactionary Catholics who blame the 'liberal' clergy, religious sisters or church bureaucrats for all of the problems of the church, both real and imagined. As the most accessible and localised ecclesiastical leadership figures, it is natural that the clergy will attract much of the criticism of those who feel dislocated and uncomfortable in the contemporary church. There is little chance that angry and discontented Catholics will be able to express their views directly to the bishop, but priests live locally and answer their own phones and front doors. In addition, as I have already mentioned, reactionaries often bypass the local church and go straight to Rome with complaints about certain 'liberal' clergy and 'heretical' teaching. There is clear evidence from overseas, especially the United States, that many officials in the Vatican spend a lot of time listening to this unrepresentative, unbalanced, and sometimes quite malicious and libellous material. Further, priests are also often criticised by mainstream and liberal Catholics because they do not measure up to the increasing expectations of a more educated, discerning and demanding Catholic community. So the clergy are caught coming and going.

It is easy to be anti-clerical, but the reality is that most Australian priests are caught in difficult, unfulfilling and isolated personal situations. Celibacy constrains many of them from normal emotional fulfilment. Despite fine rhetoric from popes and contemporary spiritual writers, the priesthood is a radically lonely existence. The very nature of their work, especially if they are reasonably successful or generous, involves priests in lives of considerable self-giving, without ever really receiving any nurture in return. While a false sense of messianic ability sustains some of them for some of the time, most mature priests eventually have to face either emotional stultification or find a creative way of dealing with the considerable ministerial demands that are made on their time and emotional energy.

Great demands are made on the most able of the clergy: they work long hours, involved with people often undergoing major life transitions (such as birth, marriage, death, grief, divorce, sickness); they are involved in counselling people seeking a sense of meaning or asking for spiritual guidance; they are expected to be preachers and able celebrants of worship. Yet these very same priests have little or no personal emotional support. It can be a depressing experience and a moment of radical loneliness to retire at night to an empty presbytery and to a silent God, after you have worked hard and have given your best to the community.

The very fact of living on the job in a presbytery owned by the church further limits the scope of the priest's personal life, so that emotionally many Australian priests are unable to think of themselves as individuals separate from the institution. As Morris West says in his new novel *Lazarus* (1990), they have become eunuchs not necessarily for the sake of the kingdom of heaven, but certainly because of it!

Central to this problem also is the church's attitude toward women. Many middle-aged and older priests have been nurtured on the seminary myth of the 'manipulative woman' who, consciously or unconsciously, is seen as a threat to the priest's commitment to celibacy. It is, of course, a form of counter-projection and, as such, is especially hard to bring into the open. It is interesting that the *Report* on the Adelaide Pastoral Team reported that there are 'grumblings about women running the diocese . . . some priests . . . are clearly not comfortable with the presence of women on the Team'. [24] Misogyny is never far under the surface in the Australian priesthood — and probably not far under the surface with Australian males generally.

Also, priests are often caught in an economic thrall. Few have any financial independence of the church, so they are dependent on the institution for board and lodging. All dioceses in Australia pay the clergy a wage that increases with the number of years of ordination, with a variation built in linked to the cost of living, with allowances for a car and other costs. But dioceses vary as to how much they say they can afford to pay their clergy. Even allowing for the fact that they receive board and lodging, priests still receive a limited salary, given that they have to pay for their own personal needs, books and seminar expenses, as well as paying for their own holidays and recreation. Very few would earn more than $10 000 per year. Given that many of the poorly paid priests are in isolated country areas

---

24. This is the team set up by Archbishop Faulkner of Adelaide that is referred to in the previous section on bishops — see p. 24.

where it is very expensive to travel, their low salary traps them in very isolated and limited communities.

## What kind of priest?

Like most professions, the priesthood today is facing increasing demands in terms of both skill and time. There is more and more work to be done by fewer and fewer ordained clergy. This is already taking considerable toll on priests in active ministry. Increasingly, the question needs to be asked: Do priests have the psychological, human and spiritual ability to cope with the demands that face them in a culture like that of Australia?

There is actually a lot of research evidence available about the psychological profile of the Catholic clergy. Priests have been far more open to scrutiny than the other professions, such as law and medicine. In the early 1970s the United States bishops' conference sponsored a study of the American clergy from several points of view, including a major psychological profile by Eugene Kennedy and Victor Heckler. There was also a major European study in the early 1970s and over the last twenty years Father Andrew Greeley has often written about the psychological development and ministerial role of priests in contemporary society. He argues that the figure and function of the priest is a central one in Catholicism and it must continue. [25]

Recently, the Catholic Institute for Ministry in Canberra produced a psychological profile of the Australian Catholic clergy entitled *A Closer Look at Australian Priests* (1990). Unlike the earlier American study, which was published publically, the Australian profile is marked 'For the personal use of the priests of Australia'. Apparently, Australian clergymen are more coy about their inner life than their American counterparts! It is surprising that such research is not publicly available to the wider church community and to those interested in the church's ministry. However, there is nothing radically new in the Australian report; it largely concurs with overseas results.

But before examining the general results of these surveys, it is worth noting that Catholic priests have two characteristics in common. Firstly, they are generally very ordinary men. As the United States survey found: 'These studies have shown the American priest to be an ordinary man, much like his fellow citizens on the scale of

---

25. For the US bishop' survey see Kennedy, E. and Heckler, V., *The Catholic Priest in the United States: Psychological Investigations*, US Catholic Conference Publications, Washington DC, 1972. See also Greeley, Andrew M., *The Catholic Myth: The Behaviour and Belief of American Catholics*, Scribners, New York, 1990, pp. 216–225.

psychological growth, and, therefore, mostly fitting into the largest category of American males of comparable background, that of emotionally under-developed males'. [26] Most priests, like most men in our society, are emotionally immature.

However, unlike most other men, celibacy constrains many priests from situations of personal intimacy in which there is an opportunity for emotional immaturity to be broken down. As Ernest Larkin and Gerard Broccolo put it:

Generally [priests] have not worked through the problems of intimacy and their level of maturity is lower than their chronological age. They do not relate deeply or closely to other people. In itself, this is not an indictment of their spirituality, because they may possibly have a high degree of theological love for their fellow men but are unable to show it . . . Psychological blocks prevent the manifestation of love. [27]

Certainly there has been a widespread realisation that lack of emotional development is a problem for many priests. The institutional church usually tries to solve this immaturity by some form of group support or community support by other priests. In Australia, the Ministry to Priests programme has been organised in many dioceses. This works on the basis of peer group support in the areas of spirituality, ministry, human friendship and fraternal companionship. While this has been helpful for a sizeable group of clergy, it obviously suffers from the inherent limitations of all peer group support: the tendency to reinforce the mores, attitudes and values of the group. The answer to emotional under-development is not necessarily to be found in groups, or even in intimate communities, but in being willing to confront the need of all human beings for intimate personal relationship and, having experienced that, to then move on to the radical aloneness built into all human existence.

While it is certainly true, as Larkin and Broccolo point out, 'psychological and spiritual growth occur best in an atmosphere of loving community . . . with one's peers', it is also true that immature males are particularly unable to form that style of nurturant community. Even in religious orders, where men are better trained for communial living, genuine growth-inducing interaction rarely

---

26. Larkin, Ernest E. and Broccolo, Gerard T., *Spiritual Renewal of the American Priesthood*, Washington DC, US Catholic Conference Publications, 1973, p. 1.
27. Larkin and Broccolo, op. cit., p. 26.

occurs. And there is something very sub-cultural, inward-looking and even narcissistic about a group of priests, in isolation from the ordinary lay community, nurturing their own spirituality and human growth. It is open to question as to how much real growth and genuine psychological development occurs in peer groups of any sort.

The evidence is that many Catholic clergy in Australia are fearful of genuine intimacy and thus remain at an immature level of development. It is significant that Broccolo and Larkin say that many priests 'excuse themselves from the pain of the growing process' and that 'The priesthood, the church and faith are used as screens and cover-ups for psychological inadequacy'. As a result it is not surprising that the longitudinal psychological surveys show that the majority of priests are to be found at the less mature end of the scale. Only a tiny percentage have reached psychological and human integration and many are caught in conflictual transition stages of human development. Presumably some priests will pass through to a higher level of maturity, some will stay stuck in conflict and unfortunately some will return to a safer, known level of development.

The immaturity of the clergy probably means that genuine encounter with God is constrained and profound feelings about God, both negative and positive, are not recognised. Thus the major source of nurture, and the peak source of experience in a self-giving, celibate existence, is cut off. The very men who purport to speak about God have, in fact, rarely if ever experienced God. Karl Rahner pointed this out years ago when he said that the priests who claimed the authority to speak about God from the pulpit are the ones who have least experienced the trancendent. This is one of the most tragic results of the lack of clerical human development. Priests have a unique opportunity to speak of the experience of God to congregations that are increasingly hungry for this type of spirituality. Cardinal Newman's motto was 'Heart speaks to heart'. But immature men are out of contact with their inner selves and the workings of God's grace within them; they are defensive about their uncertainty.

Men caught in conflictual transitions do not have a clear sense of personal selfhood and often experience tension about their identity and role. This is not a good foundation for a person who is being increasingly called upon to activate and support others in the ministry. Only a small number of priests (the various surveys put the percentage between five and eight per cent) have actually achieved a real level of human maturity. Yet one of the most genuine demands that the Catholic community makes of its priests is in the area of spiritual and ministerial leadership. Thus the question of the maturity of the human and spiritual development of the priest seems to me of pivotal importance.

Catholic priests are not alone in their human and emotional under-development. Several well-informed observers have commented to me that the situation of the Catholic clergy could be duplicated among both the Anglican and Protestant clergy. Because such clergy generally marry, the areas of under-development would probably be different, but the level of immaturity seems to be much the same. In fact, the symptoms are probably characteristic of Australian males generally. The clergy are only more extreme examples of the common problem of male emotional immaturity. Also all clergy face the even greater temptation of a false 'messianism'; that is, the unspoken and often unconscious assumption that God, the community and the world will get nowhere without them and their activity. This is a real form of self-delusion — and many clergy suffer from it!

Thus the Australian church is confronted with the whole question of the need for genuine leadership. However, the gift of leadership is usually more the exception than the rule within an hierarchical church structure. People often identify hierarchical authority with leadership. The origin of the word 'hierarchy' is in the Middle Ages. In the 11th century there was a major reform of the church by Pope St Gregory VII. He and the popes of the time tried to bring order to the church by establishing clear lines of authority centred on the papacy. The educated medieval mind tried to impose order on the world by arranging the whole of reality in a hierarchy of being — from the lowest form of matter all the way up to God. So it was natural that Pope Gregory VII emphasised the hierarchical aspect of the church.

## Qualities for leadership

Since the publication of *Mixed Blessings* I have been criticised by some because I argued that Catholicism must move from an emphasis on an hierarchical to a participative and communal model of the church. I have been accused of advocating the abandonment of church order and of any principle of church governance. The opposite is the case. I would argue that the more the church moves toward a participative model, the more it will need genuine leadership. Any old bureaucrat can double as a hierarch and can more or less affectively administer the institution. But the gift of genuine leadership will be needed if more participative models of the church emerge.

Leadership is one of those intangible realities that is found in certain individuals who are able to express through their personalities the élan, the ethos of an organisation. It is a very subtle thing — Pope John XXIII had something of it, as did Mahatma Gandhi and, in Australia, Archbishop Daniel Mannix of Melbourne. An example from our own time is Archbishop Desmond Tutu of Capetown. The true leader is one who can sum up in themselves what an organisation symbolises.

Real Christian leadership is something that transcends hierarchy. In the New Testament leadership is seen as a service. The New Testament paradigm of this is Christ washing his disciple's feet. The fundamental type of leadership needed in a more participative church is the self-effacing one of helping others to discover their gifts and then creating the right environment for people to use those gifts. Having created an atmosphere in which people can act, the genuine leader will then have to draw the whole thing together, so that the service of the church has a sense of unity and direction.

Another New Testament paradigm of leadership is provided by Peter, the man who publically denied Jesus. Because he was known in the early church community as a public sinner, he was a safe man to endow with leadership for he could never afford to take himself too seriously. If he ever degenerated into self-importance, people would only have to start talking about 'cocks crowing' and Peter would have to laugh at himself — or die of embarrassment! Only hierarchs take themselves too seriously. Because Peter was a public sinner he knew that he could not stand above others. The experience of human fragility is so often the beginning of wisdom.

I saw this vividly illustrated in the life of one of the most mature men that I have ever met. This man had been a captain in the United States navy; he had come to the priesthood late in life. Soon after ordination he had to face the fact that he was an alcoholic. For him — as for a number of other people I have met — the admission of alcoholism was the central element in an achievement of spiritual maturity. It was not just a matter of facing his alcoholism; it was a matter of embracing it. It is a weakness that will not go away; it is permanent state. The alcoholic can never say 'I'm OK now; I can stand on my own two feet'. The person who has sinned knows that at the core of human existence there is a weakness that contradicts human self-sufficiency. That person is forced to look beyond the self to others and ultimately to God.

In other words, the person who is most effective in the ministry of leadership is the person who is existentially aware that we all share the human predicament together. As Therese Hoeing says about Religious leadership:

Weakness, and the acceptance of it, is necessary to Christian leadership not only because it delineates our relationship to God, but also because it clarifies our relationship with those to whom we minister. The Christian leader can help the confused teenager, the guilt-ridden parent, or the restless octogenarian not because he has successfully dealt with each condition, but because he struggles with confusion, guilt and restlessness in his own life . . . Weakness is necessary to Christian leadership because it marks our solidarity with all that is human. [28]

Ultimately priests, like all maturing human beings, have to know and experience the weakness inherent in humanity. By accepting their own weakness those in church leadership can build a stable and coherent spirituality. The Easter liturgy's description of our human weakness as the *felix culpa* — the 'happy fault' — is so apt.

Some priests try to put off facing the *felix culpa* by playing messiah. They think that they are responsible for everything and everyone. They are over-worked because they are always out saving others; they assume that without them everything will collapse. These priests have a special need to experience the 'happy fault' in themselves and to learn that the world and the church manage to survive even without their messianic activities.

The New Zealand poet, James K. Baxter, vividly expresses what I mean by *felix culpa*; he understood it better than most. An alcoholic, who died when he was middle-aged, he was seen by many of his contemporaries as a drop-out. A convert to Catholicism, he wrote a series of letters to a priest that are real gems. In this extract he describes that weakness which is a form of strength:

My guts are consumed night and day by the knowledge of new and eminent faults of my own and the various agonies of friends. Christ is not my insulation; I suspect he is my pain. I find no escape from him whatever. He is the lion that feeds on me. Look, I could got to a beach somewhere and smoke pot and write poems and 'be happy' in the usual professional way with plays on radio and a succession of mistresses. I 'am unhappy' in a different way — somehow a sinner lost in God. He has taken away my carefully contrived insulation: the plague pit and the sky open at the same time. I know that you too are not infrequently 'unhappy'. I suspect that your weakness too is really Christ, and it must be borne with aplomb. [29]

---

28. Hoeing, Therese, 'The Necessity of Weakness for Christian Leadership' in *Review for Religious*, 39 (1980), p. 21.
29. I have the letters in an unpublished form. I am not certain if they have ever been published. If not, they ought to be.

Mature priests and mature Christians will know that Baxter is profoundly right. To know God's mercy and love, you have to experience your own weakness and sinfulness.

The great strength of the Australian clergy, however, should also be recognised. It is their common sense and generosity. Most are very hard-working and self-giving. The clergy — like most of the bishops — are often at their most helpful in one to one situations with persons in need. In these situations their ministerial instincts are often accurate and successful. The problem is that people with gifts like these are probably not the best choice for leadership in the contemporary church. A different set of gifts is needed.

There is one other great strength that the Catholic clergy have as a profession. They are usually generous to a fault and either self-critical or very much subject to the criticism of those who interact with them. In this the priesthood is unlike most of the other professions in Australia which seem to be increasingly self-engrossed.

Despite these strengths, the predominant group among the clergy, both in terms of numbers and influence, are those, who either from age or psychological profile, are committed to caution. It is not that there are no creative men left, but whether there are enough of them and whether they still have influence in the church community or are marginalised. Marginalisation occurs either through their own volition, or through a subtle conscious or unconscious process of manoeuvring them to the edges by the group with power. Creative people are often construed as threats by those with less ability. Because their minds are capable of greater abstraction and their imaginations more fertile, creative people are more likely to perceive the larger context, the bigger picture. This is often simply inconceivable to people of more limited ability, who sometimes become lost in ad hoc details and in superficial analysis.

Once cautious people become predominantly influential in a group, the possibility for change becomes progressively more remote. The critical ability to analyse and the creative ability to project possibilities for the future slowly disappear. Often the rhetoric of change and renewal is still used, but those in power lose the ability to distinguish between words and substance. A cautious leadership is unwilling to take risks and is unable to face the issues that require creative thinking. Slowly such people move progressively inward to safe parameters; they unconsciously, but happily, build the Catholic sub-culture.

I do not think that anyone should be surprised that Catholic clergy as a group are essentially conservative. The prophetic and the clerical offices are very different and are expressed through different gifts. Changing an institution is a prophetic task, a task compounded by having a clerical crisis of identity. It is understandable that priests are defensive! I think it is a miracle of grace that there are still some priests around who are willing to try the prophetic office of changing the church.

## Ministry

So far we have talked of the priest without too much reference to his primary role: ministry. There is considerable evidence of what the laity expect of the clergy. Both the report *Diocesan Synod 1989* of the Canberra-Goulburn Archdiocese and the *Report on Consultations with Laity* of the Archdiocese of Melbourne (October 1985) have set out lay people's suggestions for the future role of the priest. [30] An even more far-ranging consultation on the future of parishes was organised by the Melbourne Catholic Office for Pastoral Planning in 1990. [31]

The context addressed by both reports is usually the local parish. The 1985 Melbourne consultations are typical. There the role of the priest is sharply focused:

In the restructured parish, the priest's role would primarily be that of pastor to the people, and the one who presides at the celebration of Mass and the sacraments.

In order to achieve this, practical suggestions are made. These centre around delegation of administrative duties to parishioners or to paid administrators. The use of the word 'delegation' is significant; it seems to suggest that the priest still owns these duties which he hands over to others. Ministry, however, in whatever form, is not something that is 'delegated', but is a reality that flows directly from baptism, confirmation and church membership.

The other suggestion of these consultations focuses on the need to change lay expectations of priests: priests need time off, they need more support from parishioners, they cannot be expected to measure up the 'high expectations' of the people 'without regard for the reality

---

30. This *Report* was collated by the Melbourne Research Office for Pastoral Planning (October 1985).
31. See *Cropp News*, August 1990.

of the person'. Parishioners in turn want a voice in the appointment of their parish priest; the 'suitability of the priest for the particular parish should override the criteria of seniority in the placement of ordained ministers'.

A further area of lay concern was the shortage of clergy. A broad range of predictable suggestions emerged: prayer and encouragement of vocations, increasing the number of permanent deacons and extending the office to women, reinstating married priests and 'being open to the questions of optional celibacy and the ordination of women'. Suggestions like these have been repeated over and over again in the church in Australia and elsewhere. It is quite clear that nothing will be done about these issues until the shortage of priests reaches crisis point.

In the 1990 Melbourne consultation the responses were more precise and pointed. I will deal with those recommendations that focus on parishes later in the book. A strong theme, however, that emerged was the central role of the priest in relationship to the eucharist. The laity also showed an appreciation of the reality of contemporary priestly life. The deaneries of the northern region of the archdiocese commented:

The importance of the ordained ministry was recognised in all responses as was the essential link between priesthood and eucharist. There was a high level of gratitude for the work that was being done by the priests of the Archdiocese and a very strong sense of concern for them lest the pressures of the times lead to their being faced with an impossible work load or with loneliness arising from the fact that more and more are living alone . . . It was suggested that lay people should be involved in the appointment of priests.

This passage is typical of virtually all mainstream lay responses to the situation of the contemporary Australian priesthood. There is no trace of anti-clericalism and a realistic understanding of the situation that most priests face.

## Ordination of women

The reports of all four regions of the Archdiocese of Melbourne mention favourably the ordination of married men, the return of priests who have left the active ministry and the ordination of women. I want to focus on the question of the ordination of women. I will examine it here because it points to a deeper question, that of the meaning of ordination itself. One thing is certain: the moment these views are published the chances of my becoming a bishop (miniscule as they were) will evaporate completely, at least during Pope John Paul II's pontificate! The ordination of women is one of those questions that will get you into considerable ecclesiastical hot water; but it is also one of the questions that will not go away.

In a recent Australian book on women's ordination, *Women in the Church*, [32] Muriel Porter demonstrates that a sizeable majority in the Anglican church in Australia supports the ordination of women. But she also shows how a minority (made up of an ill-assorted alliance of many Sydney evangelicals and some Anglo-Catholic high churchmen) have used the synodal and parliamentary structures of the Anglican church to delay the majority opinion prevailing. Her book also describes the exhaustive theological investigation that preceded the debate in the General Synod.

The Anglican, Uniting and other Protestant churches have had to confront the issue of the ordination of women during the last 30 years. While it has been divisive, it has not destroyed the unity of any of the churches, although it is the cause of a painful debate within the continuing Presbyterian church. Even in the apparently conservative south island of New Zealand, an Anglican woman, Dr Penelope Jamieson, has been elected and consecrated Bishop of Dunedin. She is the second woman bishop in the world-wide Anglican communion and the first ordinary (bishop in charge) of a diocese.

Despite the scholarly debate, my view is that the question of women's ordination is not theological, but disciplinary. It is something that the church could decide to change tomorrow. It has been consistently pointed out in the Anglican and Protestant debates, and most recently by the Bishop of Edinburgh (himself an Anglo-Catholic), that the fundamental issues are not theological but socio-cultural and psychological. This is a polite way of saying that the real point is about power in the church and who has it. It is about the determination of a certain type of cleric to maintain male ecclesiastical dominance. The Bishop of Edinburgh maintains that opposition to the ordination of women is sometimes linked to immature spiritual and psycho-sexual development.

In the course of her book, Muriel Porter makes passing reference to the views of some Catholic feminist theologians. The attitude of the church forces these Catholic women to take more radical stances

---

32. Porter, Muriel, *Women in the Church: The Great Ordination Debate in Australia*, Penguin Books, Ringwood, 1989.

than their Anglican and Protestant sisters. Because women are so marginalised in the worship and life of the Catholic church — officially they cannot even be altar servers — Catholic feminists have been forced to ask a more fundamental theological question: should women seek to be ordained to the priesthood at all? Some of them go further, asking: should there even be ordination? This, of course, questions the whole present practice of the church. I think these Catholic women are pointing in the right direction: why should women want to enter a priesthood that is embedded in structures that are patriarchal and inherently clerical, especially in the way it is lived out at present in the Catholic church?

Paradoxically, Pope John Paul II is the other person who is highlighting the question of ordination. His views, of course, have nothing in common with the Catholic feminists, but oddly his actions confront the Catholic church with the same question: what is ordination and what does the priesthood mean? By demanding that only celibate males be ordained, the Bishop of Rome is, *de facto*, forcing the church to under-emphasise the eucharistic and sacramental nature of the Catholic community. This ruling places a large proportion of the Catholic world (in Latin America, Europe and increasingly in the Anglo-American world) in a priestless situation, thereby forcing a change in the traditional nature of the Catholic church, which is intrinsically sacramental. By linking celibacy to the priesthood, John Paul II prevents the ordination of suitable people and thus imposes on many communities a Sunday bible and communion service, led by an unordained person, in place of the normal eucharistic celebration.

Ironically, while it is obviously not his intention, his action seems to me to force the church to ask the question: what is the meaning of ordination? Do we really need an ordained priest? Can Catholics make do with a Sunday bible and communion service? And a further question also needs to be asked: how far can this denial of a regular eucharistic celebration be pushed without denying an essential element of the nature of the church? The practical denial of the eucharist is apparently being allowed by the pope to preserve celibacy, clericalism and Roman power over the local churches. The Australian Catholic church has not yet fully experienced the acute consequences of a shortage of clergy, although there are already about fifty parishes without a resident priest. We are only on the edge of priestless parishes, largely because of the committed work habits of most of the Australian clergy.

Sister Ruth Egar of Adelaide recently reported on the views

of a number of local priests about their lives, beliefs, values and spirituality. [33] She says that some priests see themselves as men who are willing to 'be there' for people on any and every occasion, and she quotes one priest as saying, 'I don't know any priest who wouldn't do anything for anybody who asked'. I know that there is a tendancy to idealise one's profession, but I think that this is fairly typical of the work patterns of many of the Australian clergy.

But to return to the question of ordination. The pope and the feminist theologians, both from radically different points of view, confront the church with the question of the meaning of ordination. Is the ordained ministry exclusively linked to the clerical priesthood as we know it? And is the priesthood necessarily linked to the liturgical and sacramental ministry? Even now lay people are the ministers of the sacrament of marriage (the priest is only the witness), they can baptise in case of necessity, and Saint Thomas Aquinas suggests that lay people may be able to reconcile each other in an 'incomplete sacrament' when no priest is available. [34] As lay people increasingly participate in more and more of the church's ministry, the ordained clergy are increasingly reduced to Mass celebrants.

I am not suggesting that we abandon ordination as a sacrament, but I do think that we are at a point in the development of the church where the meaning of it will slowly change and expand beyond its present implications. There is no theological reason why this should not happen. The sacraments have always been under the control of the church. They have evolved in meaning and practice in the course of the history of the church and some of them, such as penance or reconciliation, have undergone radical change in form and significance. [35] So, in fact, has the sacrament of ordination. Father Kenan Osborne's study of the history of ordained ministry points out that even the word 'ordination' itself creates problems:

Ordination, in the sense that one finds it in the third century onward, is not at all visible in the New Testament. This does not mean that the later, elaborate ordination rituals run counter to the New Testatment data, but

---

33. Sister Ruth Egar's articles are in *Shepherd's Pie: A Newsletter for Priests in the Archdiocese of Adelaide*, November 1989, February and April 1990.
34. Dallen, James: *The Reconciling Community: The Rite of Penance*, New York, Pueblo, 1986, p. 188.
35. See Dallen, op. cit., pp. 5–201 for a history of the sacrament of penance.

it does mean ordination, whether for men or for women, is not clearly attested to in the New Testament itself. [36]

Also, the latin word *sacerdos* — 'priest' — only began to be applied to bishops around 200 AD, and to those we call 'priests' between 400 and 500 AD. [37]

Because the sacraments are part of the organic development of the church, contemporary Catholics also can participate in the evolution of the sacrament of orders. What the contemporary church might develop over time is a situation similar to that of the early church. No longer would ordination be seen as a change of status from laity to clergy, but it would become an anointing of men and women 'set aside' for the various ministries of the church. [38] This implies that the sacrament of ordination will expand in meaning. It will be oriented fundamentally to ministry and its practice will be determined by the needs and priorities of the local church. Ordination could come to mean the anointing of chosen people, who after specific training for a particular ministerial function are called and missioned for that ministry in a ceremony akin to today's ordination.

Several things need emphasising here. Firstly, people need to be chosen and trained for ministry in accordance with specific gifts which the church discerns they have. There would be a range of gifts and corresponding ministries, the categories varying from place to place according to need. And I stress that this would not just be a matter of 'discernment' of specific personal gifts; it would also involve formation and training. God save us from self(or other)-appointed amateurs! Someone, for instance, may be excellent in working with the sick. They would then be trained as perhaps, a hospital chaplain or as a minister working with the sick or dying at home. They would then be called, anointed for this ministry and empowered to administer all the sacraments associated with the ministry — taking communion to the sick, personal counselling and reconciliation, administering the sacrament of anointing.

---

36. Osborne, Kenan B., *Priesthood: A History of the Ordained Ministry in the Roman Catholic Church*, Paulist, New York, 1988, p. 81.
37. Osborne, op. cit., p. 160.
38. Edward Schillebeeckx points out that the Latin word *ordinatio* originally meant entry into a particular social *ordo* or 'class' of the late Roman social world. In our terms, then, ordination means a change of 'class' from laity to clergy. See *Ministry: Leadership in the Community of Jesus Christ*, Crossroad, New York, 1981, p. 39.

Secondly, there will be some ministries that will be operative in all local churches: for instance, the tasks of celebrating the eucharist and worship, the preaching of the word and teaching catechesis, the ministry of leadership, and those gifts oriented to the unity of the local church. These will probably require a corps of full-time ministers, who may or may not be committed for life. People will obviously move in and out of their ministry according to circumstances. Family commitments may well be one of the key factors determining this participation.

Thirdly, the local church and its needs will be the basis upon which the ministerial structure will be built. In that case, the church in rural Australia might be different in its structure from the church in the large cities.

All of this has seemingly taken us a long way from the ordination of women. But I return to my original assertion: the real question is not the ordination of women, but of the meaning of ordination itself. Once you open up that question, you cannot be sure where it will take you. The discussion has also apparently taken us a long way from the present situation of the Australian Catholic priesthood. However, the danger is that priests will remain bogged down in uncreative situations if no one tries to imagine a possible future. Human beings need some plausible models toward which they can work. There is no future without imagination.

## Towards a new identity

The Australian priesthood, then, is a profession caught up in a crisis that could result in a profound shift towards a new identity in the church; or it could involve a retreat into a Catholic sub-culture in which the church would simply be a reactionary rump. A considerable minority of priests are working toward creative change and they are willing to integrate lay people more and more into the ministry of the church. But this is not easy, and the question of power and who has it is still an important but unarticulated issue in the process. Most priests, even quite imaginative ones, are threatened by the thought of the laity taking over more and more of 'their' ministry. Will anything be left for them? Will they have any real power? As one priest said to Sister Ruth Egar:

I think we're still in a subtle power struggle, a search for control. 'The people are the church' statements can be felt as 'The people want to control you'. If I'm honest, I'm threatened by it. We're not yet giving freedom to each other's spirituality.

I think that this comment is very significant and it is rare that it is articulated so honestly. Lay people should be very clear that they are asking for power and that they are asking the clergy to surrender it. No one surrenders power easily, especially when power and personal identity are as closely intertwined as they are in the priesthood.

Unfortunately, however, some Australian priests are not open at all, and have not yet even reached the point of allowing the laity any say, let alone any leadership role, in the ministry of the church. These men have retired into the priesthood as a clerical profession and they guard their prerogatives and privileges as closely as any conservative lawyer or doctor. Men of this ilk see the priesthood as a closed club. They even joke about belonging to the 'most exclusive' club in the world.

Today many priests seem to shelter behind a bureaucratic attitude; they are involved increasingly in becoming organisers of the day-to-day activities of the parish. Sadly, many priests have retired into the Catholic sub-culture where they are insulated from new and difficult questions and where their role is safe from challenge. An example of this is the way in which some priests hide behind church law and bureaucracy in order to avoid making hard pastoral decisions. People are told 'I can't do this or that, because it is against the law of the church', or 'The bishop does not allow it' — sometimes even when this is not true. Now, I am not an anarchist, or even an antinomian, but traditionally there has always been within Catholicism a sense of what is called 'pastoral discernment'. There are times when you would never think so.

An American priest told me recently that his bishop exercises what he calls 'pastoral cruelty' rather than 'pastoral care', by creating a bureaucratic obstacle course for even the most ordinary requests.

A common example of such a pastoral obstacle course is the promises that the Catholic spouse has to make before a 'mixed marriage' — a marriage in which only one party is Catholic. The Catholic has to promise that he or she will do all 'in their power' to have the children baptised and brought up as Catholics. Now this can create real problems for the non-Catholic party, especially if this person is committed to his or her own faith. The irony is that it is often those with little religious conviction who have the easiest time getting the 'dispensation' required; it is easy to allow your spouse to make promises when you are not deeply convinced about religion yourself. It is much more difficult if your faith is important to you.

Perhaps the greatest opportunity in the present confusion about the priesthood is that it is a *kairos*, a moment of vulnerability, opportunity and choice. Throughout the history of the church the priesthood has evolved and changed. Right now we again have a marvellous opportunity for creative change. The United States bishops express this clearly:

Priestly ministry is not a finished reality, fully achieved, like a work of art. Neither is it something frail, a fragile object, unable to withstand the taxing passage through change and time. Rather, what the sources of faith and theology reveal is a priestly ministry as a living reality. [39]

But as most priests who have tried to change ecclesiastical and pastoral attitudes and practices know, creative change is easier said than done!

In some ways this long discussion of priesthood and ministry has seemingly bogged us down in an analysis of the very sub-culture that I have previously criticised. To an extent, this is true. But to neglect the priesthood is to neglect the most important leadership cadre in the church. As long as Australian priests are caught up in a struggle about their identity and role, or see themselves solely as the clerical and bureaucratic caste who supervises the church at the local level, they will never have the vision to lead the Catholic community into the future. The safest place for people with an identity crisis, or a bureaucratic-clericalist mentality, is a sub-culture.

If the church is to break out of the sub-cultural trap there will have to be some priests who 'seize the moment'. And there are a number who already have already achieved significant change in Australia. Some are well-known, others known only to the people whose lives and communities have been affected. The most effective priests often work in an unassuming, quiet way that gets little public notice. Examples of priests who have seized the moment and built something new are men like Fathers Ted Kennedy, Michael Fallon and Peter McGrath, and, perhaps less known, Fathers Noel Molloy, Brian Stoney and Barry Moran. This list is not, of course, exhaustive.

Ted Kennedy has been committed to Aborigines in Sydney's

---

39. US Bishops' statement, *As One Who Serves*, c.2.

Redfern for many years. Michael Fallon was the centre of a community of young lay people when he was chaplain at the University of New South Wales. He has gone on to become one of the most successful teachers and writers on the New Testament; his work is especially directed to the laity in order to give them the biblical knowledge needed for them to assume leadership and deepen their spiritual lives. Sixteen years ago Peter McGrath began the Passionist Family Group Movement in Terrey Hills, an outer Sydney suburb. The movement is now spread across 40 parishes in Australia. The parish was begun in 1972 and it quickly grew, 'largely because of the leadership style adopted by its pastors'. [40] Peter McGrath has also been provincial of the Passionists in Australia.

Noel Molloy worked originally in the Sydney Archdiocese educating parishioners in community formation. He put his own teaching into practice in the rural parishes of Bordertown and Yorketown in South Australia. Brian Stoney is a Jesuit. Over the years he has shown a deep commitment to working with people on the streets. He works with a group of lay people and has had an extraordinary effect on many of them. He worked at Corpus Christi Community in Melbourne for homeless alcoholics and now is in King's Cross in Sydney. [41] Barry Moran is one of those priests devoted especially to the often neglected outer suburbs of our large cities. He has worked in three outer Melbourne suburbs since 1974 — Keysborough, Frankston and now Werribee. He has been described by a lay colleague as having 'an instinct for the suburban'. This is borne out in an article he wrote for *The Catholic Worker*:

It needs to be said that a Melbourne suburb is a legitimate place to be a Christian. A suburb is not synonymous with sin. There are times when I think a lot of people want us to feel guilty for living in the suburbs, in a detached house with a mortgage . . . It is an awful thing to have a job, a spouse, children and a house, and say, 'Well, here I am and I still have to make sense of this bit of living because my job is not exciting. It is something that ought to give me a sense of who I am and what I can do, but every now and then I need to stop and think what does it all mean, and what does it have to do with me that I am concerned with what I saw on TV about Ethiopia'. [42]

---

40. Britt, Mary E., *In Search of New Wine Skins: An Exploration of Models of Christian Community*, Collins Dove, Blackburn, 1988, p. 48.
41. A video about Corpus Christi Community has been made called *Dance in Darkness*. It is available from Catholic Communications, 188 Brunswick St, Fitzroy, 3065.
42. *The Catholic Worker*, May–June, 1985, p. 6.

The one thing that all of these priests have in common is a commitment to empower other people, especially the laity. They are leaders in the truest sense, acting as bridges from the clerical/ hierarchical model of church to the emerging community model. It is a pity that there not more like them.

# *The laity*

It is very difficult to generalise about 4 064 413 people. [43] This is made even more difficult by the fact that about 45% of them were born outside Australia, or are children with at least one parent born overseas. The Catholic church is, by far, the most multicultural of the churches. But despite the difficulty of generalising, it is important to try.

## An Anglo-Australian church?

Let us start with the first group that established the Catholic tradition in Australia since white settlement, the Irish-Australians. Actually, from the middle of the 19th century, the church was never totally Irish; there were small groups of Italian, German and later even Lebanese Catholics in Australia. The first Lebanese priests, who belonged to the Maronite, not the Latin rite, arrived in Sydney in 1889. [44] But the reality, or perhaps, more accurately, the myth of Irish-Australia dominated the Catholic church until 1947. That year saw the advent of post-World War II European migration. As Patrick O'Farrell has shown, the Irish influence in the church had faded by 1940 or even earlier, but, as he says, its 'influence remained in an Irish style of religion — clerical, authoritarian, non-intellectual: it was Australian Catholicism, with a brogue'. [45]

The 'Australian-ness' of the church was shown by the fact that Catholics were only too happy to join the national preoccupation with material comfort. They struggled out of the working class and by the 1960s Catholics had 'arrived' in Australia.

Yet, paradoxically, Australian Catholics seemed able to maintain a sense of separate existence for most of this century. This was probably very closely connected with the struggle to support the

---

43. See Australian Bureau of Statistics, *Australia in Profile*, p. 19.
44. Campion, Edmund, *Australian Catholics*, Viking, Melbourne, 1987, p. 181.
45. O'Farrell, Patrick, *The Irish in Australia*, University of NSW Press, Kensington, 1986, p. 294.

Catholic education system before the granting of state aid in the 1960s. It was probably also linked to the fairly tight social control the clergy were able to maintain, through preaching and the confessional, over the sexual and marital lives of lay Catholics. In this way the illusion was fostered of a Catholic 'ghetto', but it probably never really existed in any genuine racial sense. Since the advent of state aid, Catholic schools have become more and more like their state counterparts and the emphasis on ecumenism since Vatican II has meant that Catholic separatism has been broken down in favour of a more ecumenical, common Christianity. As a result Anglo-Australian Catholics have become religiously more and more like their Anglican and Protestant counterparts.

The recent reports from the Christian Research Association by 'Tricia Blombery and Philip Hughes tell us much about the beliefs and attitudes of Anglo-Australian Christians. [46] The reports show that the majority of practising Christian laity in Australia place a priority on religious subjectivity and a personal experience of God. They want a church that will serve as a source of peace and refuge from the stresses of the secular world. This, no doubt, helps to explain why only a minority of Catholics show a serious commitment to papal and episcopal teaching on social justice. Historically Anglo-Australian Catholics have had a strong commitment to social justice but only because it was to their own advantage. Migrant Catholics — like migrants generally — are usually too concerned with establishing themselves in the new country to be socially concerned with other people's problems.

Most Catholics want priests and church leaders who comfort them like a friend. They tend to ignore those who challenge them to think and act prophetically. The church as sub-culture should be a great place for people like these! On a more positive note, there has been a retreat over the last two decades from a legalistic religiosity that focuses on God as judge, to a subjective and experiential religiosity that focuses on God as a friend and companion. I do not think, however, that the Catholic church can accept this subjective religious style in the long term, for it is only a very partial realisation of the reality of Christian faith, and ignores the fact that social justice is

---

46. Hughes, Philip J. & Blombery, 'Tricia, *Patterns of Faith in Australian Churches: Report from the Combined Churches Survey for Faith and Mission*, Christian Research Association, Melbourne, 1990. See also Hughes, Philip J., *The Australian Clergy*, Christian Research Association, Melbourne, 1989.

an integral part of living Catholicism. Only a minority of Australian Catholics have ever shown a strong commitment to social justice.

Over the last twenty-five years Anglo-Australian Catholics have become more verbal, cerebral and, in a way, more 'Protestant'. One of the great failures of contemporary Catholicism is a partial loss of the sense of the visual and symbolic and, at a deeper level, a loss of the sense of the sacramental. The church has retreated from genuine ritual to a verbal, 'talkative' worship that is truly boring! A lot of Catholics say that they want to 'experience' God. However, they understand experience in a quite superficial sense referring to a vague and comforting feeling of God's presence which makes no real personal demands.

Many Catholics claim that they are unable to achieve this comforting feeling in contemporary worship. They retreat into a spirituality that often has more in common with pop psychology than genuine Catholic spiritual tradition. It is easy enough to understand the reason for this retreat into the subjective. In times of shifting values many seek security in a God who comforts them. This is especially true at the present moment as the Catholic church passes through a period of profound change and dislocation (what I referred to in *Mixed Blessings* as a historical 'mutation'). Personally, I see this emphasis on the subjective as a transition stage rather than a permanent state.

Historian Edmund Campion argues that Australian Catholicism has also become increasingly conservative, as Catholics have moved from working class suburbs to middle and upper middle class areas, since the Second World War. [47]

The conservatising of Australian Catholics had been proceeding apace since World War II. By 1980 it was apparent in the high visibility of Catholics in both the Liberal and National/Country parties. It was also apparent in the knighting of half a dozen Archbishops — earlier generations of Australians had despised such imperial honours as anti-Australian. It was very apparent in responses to the Vietnam War, the most troubling national moral question in recent decades.

Campion's assessment is right; many establishment Catholics have forgotten — if they ever knew — the sense of social and communal interdependence that is at the heart of Catholic social teaching.

The decade of the 1980s was a period of self-indulgence when both

---

47. Campion, op. cit., pp. 239–240.

major political parties succumbed to the blandishments of so-called 'economic rationalism', which is just another name for selfish and amoral capitalism. It is striking how many Catholic politicians and entrepreneurs were prominent during this decade. This period of self-indulgence began dying a slow death with recent revelations of criminal and irresponsible financial activity that reached to the highest echelons of a number of Australian governments. The reappraisal of public morality forced by these events, together with the crisis in the Gulf demands a whole new approach for the decade of the 1990s. Thus I return to my theme: the opportunity is there for Catholicism to offer some leadership in this society. The Catholic community faces a real *kairos*.

While Catholic and Protestant Anglo-Australians have become more and more alike, this is not true for Catholics from non-Anglo-Australian backgrounds. Southern European, Latin American and, to an extent, Asian Catholicism is more cultic and tactile. Theirs is a faith that is deeply embedded in ritualistic and sacramental life-transitions. It is less individualistic and more oriented to family and clan. The experience of God is more linked to dramatic community ritual, to feasts and images of Our Lady and the saints. It is concerned with feeling; but it is the feeling that arises spontaneously from communal religious fervour. It is not a purely subjective experience. In explaining the Italian approach to Catholicism, Franco Cavarra describes an interesting contrast:

The Australian Catholic tradition is characterised by a no-nonsense, no-frills, joyless attitude to religion. Religion as a painful duty. Italians demand and insist that their experience of the infinite move them emotionally. They want to be mesmerised by the oratory of a preacher who will take them on flights of spiritual fancy, they want all the colour and pageantry of religious processions of their favorite saints and the Madonna. Motherhood holds such a special place in Italian life that the incarnation can only be understood in terms of God and a human mother. [48]

While linked to family and clan, non-Anglo Catholicism still does not have a sense of society as a communal structure; it thus finds the concept of social justice abstract and irrevelant.

Eastern European Catholicism, in contrast, is often intimately linked to a sense of national history and culture and is often strongly

---

48. Cavarra, Franco, 'Bread and Roses: Italian Catholics in Australia', *The Catholic Worker*, May–June 1985, p. 5.

Marian in emphasis. For the Poles, for instance, devotion to Our Lady of Czestochowa is central both to faith and to the national myth. For many Eastern Europeans the church is an integral part of their national identity. [49]

Asian Catholicism is different again. Like that of the southern Europeans it is more tactile, festive and devotional. The figure of the priest is a very important focus in the life of the community, especially among the Vietnamese. This is also true for Italian Catholics. For such people, the priest is the symbol of a divine 'presence'. It would be difficult for them to think of the church with 'priestless parishes' as an acceptable alternative.

These ethnic Australian Catholics stand in interesting contrast to their Anglo co-religionists. We know a lot less about them, because, being migrants, they are much less likely to have participated in surveys about religion. They are the neglected group within Australian Catholicism which, with the exception of Catholic educational authorities, still directs much of its energy to the needs and interests of Anglo-Australian Catholics. Also, many of them have been placed under considerable pressure by parish priests to conform to the religious priorities and parish structures of the Anglo-Australian majority.

## Who is a Catholic?

I have said it is difficult to make generalisations about four million Australian Catholics, but I must now endeavour to do exactly that. Who and what is a Catholic? A broad definition is required, for Catholicism is an inclusive rather than an exclusive religion. I would say a Catholic is a person who is baptised as a Catholic and who self-identifies as such. I think that such a broad definition of Catholicism is appropriate for a religion that has, historically, avoided sectarianism and which, theologically and sacramentally, places much more emphasis on God's fidelity than on the frailty of human faith.

Catholics can be divided into four general groups, which can then be further subdivided. The first group consists of those who have given up all Catholic identification, except for national census purposes. There are others, of course, who have given up even that. The reasons for the abandonment of the church are, no doubt, many and varied. Given its marginality, I will not consider this group here

---

49. McKay, Jim & Lewins, Frank, 'Religious conflict and integration among some ethnic Christian groups', in Black, op. cit., pp. 171–172.

— although these people are sometimes deeply religious and the investigation of the reasons why they abandoned the church might well be very instructive for the Catholic community. Some of these people, indeed, will want to have their children baptised and later enrolled in Catholic schools.

The second group are those I will call 'cultural Catholics'. The third group comprises committed Catholics whose religious practice is irregular — the so-called Christmas and Easter Catholics. The fourth, and most important group, are those with serious beliefs and a commitment to regular liturgical and sacramental practice. I will examine each of these three groups in turn.

Firstly, the 'cultural Catholics': the term was coined by Gerard Henderson. [50] Henderson is a newpaper feature writer who, for a time, was a senior advisor to the then leader of the Liberal Party, John Howard. He is also the author of *Mr Santamaria and the Bishops*, a study of B.A. Santamaria's influence on Catholic action and the formulation of early episcopal statements on social justice. [51] Many 'cultural Catholics' are prominent in politics, literature, the professions (especially law), business and the media. They are generally not practising Catholics and their interest in the church is often largely nostalgic but they publicly acknowledge the ethos that their Catholic education provided for them. Most cultural Catholics, it seems, do not necessarily believe in the church's corpus of teaching and are apparently not strongly religious. Their attachment to Catholicism is historical and cultural. They often want their children baptised, but not necessarily enrolled in Catholic schools.

However, a note of bitterness toward the church can be discerned in some of these cultural Catholics. They are like people dealing with unresolved anger and loss. Some of them seem nostalgic for an Irish-Australian past that is largely the product of their own imagination and never really existed. One identifiable strain of cultural Catholics in Sydney is closely linked to the right wing of the New South Wales Labor Party. Another strain espouses economic rationalism and various right wing causes and is loosely linked to the Liberal Party. Many cultural Catholics are socially and economically successful and seem to despise middle class, suburban Catholics.

---

50. See Campion, op. cit., p. 247.
51. Henderson, Gerard, *Mr Santamaria and the Bishops*, Studies in the Christian Movement, St Patrick's College, Manly, 1982.

The neo-conservatives among cultural Catholics often presume to advise and criticise those who are still active in the church. They bemoan the introduction of the English liturgy, the prevalence of 'liberal' clerics and 'left-wing' ideas (in other words, the church's social justice teaching). But their absence from the active Catholic community means that they speak largely as outsiders, and their opinions often betray them as quite ill-informed outsiders at that. In my view such 'cultural Catholics' have little to contribute to the future of the Catholic church in this country. They have simply excluded themselves.

The third group practises their Catholicism irregularly. But their Catholic faith is obviously important to them and they want to hand it on to their children. They want their children to be baptised and attend Catholic schools. They are what are often unfairly called 'Christmas and Easter Catholics'. Generally, they do not think much about their faith, but it is an obviously important enough element in their lives for them to want to hand it on to their children. They see religion as a fundamental force that sanctifies the major life transitions of both the individual and the family. A sizeable group of ethnic Catholics would fit into this category. Such people see the church as a means of attaining personal identification with God and also as a means of enhancing and sanctifying family and social bonds.

Sadly, there have been times when I have seen over-zealous priests demand 'conversion and commitment' from Catholics in this category. These priests are often simply showing their own self-righteousness and lack of insight. They demand more commitment than these people can give and show little understanding of their mentality. Their attitude may be changed by a more subtle approach to evangelization: confrontation only causes pain, confusion and sometimes a wasteful separation from the church.

The fourth group is Catholics whose practice is regular and whose convictions are deep. The future of the Catholic church in Australia in the last decade of the 20th century lies with this group. Committed Catholics range across a spectrum from those who practise regularly because they consider that 'it is a sin not to' (in other words those whose commitment is basically moralistic; these days there are fewer and fewer people like this), to those with a fully articulated faith that expresses itself in commitment to a genuine spiritual life and ministry to others. For these Catholics active faith forms the core element and the basic meaning structure of their lives.

Probably less than about twenty-five per cent of Catholics could be classed within this 'serious' spectrum and they range in religious attitude from conservative to liberal. These are people who have come to an experiential and living faith which they are consciously determined to live out within the context of the Catholic church. The sacraments — especially the eucharist — are at the core of their lives, and the future of the Catholic church and community is important to them; Catholicism provides them with their basic meaning structure. The Australian church must hold on to this group of people if it is to have a future, for these are the people who will form the ministerial core of the future.

## The laity's contribution

Catholic lay action has a long history in Australia and elements of this history are quite instructive for the contemporary church. Edmund Campion has argued that the Catholic church in Australia was founded by Irish convict laity. [52] His view is that these laity formed something akin to 'base communities' that met for informal prayer and worship between 1788 and 1820, the years that preceded the arrival of the first permanent priests. The implication here is that clergy were not really necessary to build a Catholic community. 'Almost uniquely in the history of world Catholicism, Australia was not founded by bishops . . . nor by priests but by the laity — and convict laity at that'. [53] In my opinion, this view is historically incorrect. I would argue that Australian Catholicism was not formed until the period from 1830 onwards. This was the time when an assertive, professionally-trained clergy began to exercise real social and religious control over the personal lives of Catholics. A parallel development can be seen in the other Christian churches. [54]

However, as the Catholic church developed through the 19th century, various attempts were made to encourage lay participation in the life of the church. Several of these attempts centred around Father John McEncroe, one of the few truly great priests of Australian Catholic history. But these attempts were largely foiled in 1859 through the determined opposition of Archbishop John Bede

---

52. Campion, Edmund, 'John Joseph Therry in 1988', in Brown, M. & Press, M., *Faith and Culture: A Pastoral Perspective*, St Patrick's College, Manly, 1984.
53. Campion, op. cit., p. 3.
54. I have argued this in detail in my Ph.D. thesis *William Bernard Ullathorne and the Foundation of Australian Catholicism 1815–1840*, Australian National University, Canberra, 1989. See especially pp. 147–150.

Polding. [55] He maintained a strict level of episcopal authority. After Polding's death, and the Irish clerical takeover of the Catholic church led by Cardinal Patrick Francis Moran, lay activity was smothered by clerical dominance.

This dominance lasted for fifty years. Lay participation in the life and ministry of the church really began in the modern sense in the 1930s. The origins of lay Catholic action lie in 1931 in Melbourne, with the foundation of the Campion Society, a group of laymen associated with Melbourne University, who believed strongly that Catholicism had something to contribute to contemporary culture and that Catholic laity should act in the world to achieve this. Colin Jory, historian of the Campion Society, explains it this way:

Where the Catholicism of those around them tended to be defensive, theirs had acquired an assertive, Chestertonian quality. Where others interpreted the Catholic faith intellectually in terms of traditional apologetics, they did so also in reference to its contribution to Western civilization and culture. They thought of the church not as a static entity, but as a dynamic force in a desperately confused world. [56]

Ironically, it was an Irish cleric, Archbishop Daniel Mannix, who really encouraged the Campions. But then Mannix was always a different type of Irishman from the rest of the Irish-Australian bishops.

The work of the Campion Society is also closely related to the 1937–38 foundation of the Australian National Secretariat of Catholic Action. [57] After the bishops approved the foundation of a National Secretariat in 1937, a division emerged between the Sydney and Melbourne brands of Catholic action. The assertive Melbourne-based Campions wanted lay action to be organised on the basis of the model evolved by the Belgian priest, Canon (later Cardinal) Joseph Cardijn. It was called the Jeunesse Ouvrière Chrétienne (JOC) — in English, the Young Christian Workers (YCW). It was the JOC that developed the 'see, judge, act' model of ministerial action — see the reality around you, especially in your workplace, judge what is happening

---

55. For an account of crisis of 1859 see Hosie, John, '1859, Year of Crisis in the Australian Church', in the *Journal of Religious History* 7 (1963), pp. 342–361. See also O'Donoghue, Frances, *The Bishop of Botany Bay: The Life of John Bede Polding, Australia's First Catholic Archbishop*, Angus and Robertson, Sydney, 1982, pp. 111–119.
56. Jory, Colin H., *The Campion Society and Catholic Social Militancy in Australia 1929–1939*, Harpham, Sydney, 1986, p. 34.
57. Jory, op. cit., pp. 88–91.

in the light of the bible and your discussion with fellow Christians in the movement, and then act ministerially in the real world. The JOC was lay-led and it took Catholic action into the modern industrial milieu.

In Sydney, Archbishop (later Cardinal) Norman Gilroy took a far more conservative approach and wanted parish-based lay groups to be set up under tight clerical control. The Sydney model was much more narrowly ecclesiastical and more oriented to social activities for young people. Outside Sydney, the Melbourne model prevailed and the Australian National Secretariat of Catholic Action was set up (in Melbourne) by the bishops. By the time of the outbreak of the Second World War, the focus of the National Secretariat had shifted from an emphasis on lay intellectual and ministerial action to the dangers of communism, especially in the Australian trade union movement. By the 1940s Catholic action was largely under the influence of B.A. Santamaria. It was he who established, with the support of Archbishop Mannix, the Catholic Social Studies Movement (the 'Movement') and the industrial groups (thus the term 'groupers') as instruments to destroy Communist influence, principally in the trade unions.

The later tragic and bitter divisions within the Australian church (especially between Sydney and Melbourne) over Catholic action, and the Labor Party split in the 1950s and early 1960s are beyond the scope of this book, but the results of these divisions still deeply afflict the Catholic community. The ideals of these early lay activists were originally ahead of their time, but tragically, they became bogged down and distorted by party politics.

But not all Catholic laity followed the Santamaria interpretation of events. There was a persistent group of laity in Melbourne from the 1950s onward which centred around the independent lay journal, *The Catholic Worker* (ironically first edited by Santamaria himself), which opposed both the Vietnam War and conscription and supported the reforms of the Second Vatican Council. This group was largely made up of academics and well-educated Catholic laity, such as poet and university lecturer, Vincent Buckley (who left the group because of his support of Australian involvement in Vietnam) and philosopher, Professor Max Charlesworth. [58] Patrick O'Farrell describes the origins of this intellectual ferment:

---

58. However, Buckley has indicated that he later regretted his support of the Vietnam War, which he eventually came to see as 'unjust'; see his autobiography *Cutting Green Hay*, Penguin Books, Ringwood, 1983, pp. 163, 166.

The active beginnings of this new surge of intellectual life go back to the early 1950s. As before, Melbourne led the way. In 1950 a Mannix scholarship was established to send Catholic university students overseas for further studies . . . its recipients tended to select as places to study secular universities or the avant-garde Catholic, not the traditional institutions of Catholic learning. [59]

It is significant that Max Charlesworth was one of the recipients of the Mannix Scholarship. O'Farrell characterises these intellectuals as 'boisterously critical'. He also notes the parallel emergence of a traditional Thomist intellectual tradition, beginning in the 1940s and centred on the Aquinas Academy in Sydney.

Catholic traditionalism is still an important influence in the contemporary Australian church. Its history is, however, discontinuous; it has ebbed and flowed, being represented by different movements at different times. Paralleling traditionalist intellectual currents, but independent of them, is the continuing influence of B.A. Santamaria in the life of the church, principally in Melbourne, and lately with an increasing national audience through the magazine *AD 2000*.

Until it ceased publication in 1976, *The Catholic Worker* was also a vehicle for the ideas of lay people who wanted to play a larger part in the ministry of the church in the period during and just after Vatican II. *The Catholic Worker* had lost its wide-spread influence after it was no longer allowed to be sold in Melbourne churches in 1955. Its title and spirit was revived by another group of laity in Melbourne in the 1980s, but economic and logistical difficulties soon forced the second *Catholic Worker* to cease publication. A number of small lay-supported publications started in the 1960s and 1970s, but most did not survive beyond a few years. It has only been those supported by the religious orders, such as *Compass Theology Review* (published by the Missionaries of the Sacred Heart), that have survived. *Compass* has now been published for 25 years.

What the two *Catholic Workers* and the other lay-supported publications have done is to signal the emergence of a new style of lay leadership in the church that has developed over the last thirty years. This leadership has emerged from a number of sources: those dissatisfied with the Santamaria/Movement interpretation of events in both Australia and the wider world; those influenced by the new, more open spirit that has swept across the church since Vatican II; and those influenced by the way in which the Vietnam War forced

---

59. O'Farrell, op. cit., pp. 408–413.

people to make moral decisions about Australia's role in that conflict.

Typical of this group of lay people is Professor Max Charlesworth who has, for many years, articulated an intelligent and coherent Catholic response to contemporary issues. He has been much more effective in presenting a Catholic perspective to the wider Australian society than many of the church's professional theologians. Educated at Melbourne University and at the Catholic University of Leuven, he and his wife were instrumental in bringing the French-founded Équipes de Notre Dame (Teams of Our Lady) to Australia, one of the first attempts to develop a coherent theology and spirituality of marriage, through bringing small groups of couples together for intense discussion, prayer and worship. This was paralleled, at a more popular level, by an older group, the Christian Family Movement (CFM), founded in Chicago. The CFM used a basic group approach to lay formation: small groups of married couples met to discuss and deepen their faith. It was less intellectual and less structured than the Teams of Our Lady. The CFM flourished in Australia in the late 1960s and early 1970s. CFM members, lawyer Patrick Crowley and his wife Patty, were appointed to Pope Paul VI's ill-fated birth control commission after Vatican II. A consultation about the successes and failures of the rhythm method among CFM members was influential in the birth control commission's deliberations. [60]

Mention of the CFM and the Teams opens up discussion of a number of other movements that have been important in involving lay people in the church over the last thirty years. Many couples from the CFM were later absorbed by Marriage Encounter, which also came to Australia from the United States in the mid 1970s, although its origins were Hispanic. It was brought to Australia by a Sydney couple, Mavis and Dr Ron Pirola. The Pirolas, through Marriage Encounter, have been very influential on the direction taken by married and family spirituality in Australia over the last twenty years. It is experiential and personalist in orientation and focused almost entirely on the relationship of the individual couple. Its philosophy is that through a good marital relationship, partners develop and deepen a relational spirituality, become better parents and more deeply committed people within the local parish. Parallel to this was the development of Cursillo, another American movement with Hispanic origins. The youth movement, Antioch, has evolved out of Marriage Encounter.

---

60. Kaiser, Robert Blair, *The Politics of Sex and Religion: A Case History in the Development of Doctrine 1962–1984*, Leaven Press, Kansas City, 1985, pp. 74–76; 78ff.

All of these more contemporary movements are evangelical, devotional and personalist in orientation, but what they have done is to give a considerable number of Catholics an enthusiasm and an experience of faith which can then develop and be directed outward toward ministry. As Saint Paul says, 'I believed, and therefore I spoke'. Belief, in other words, struggles toward ministerial expression. It is this ministerial energy that needs to be tapped and channelled for the church to move forward.

Of course, not all serious Catholics have been part of these enthusiastic movements. Committed Catholics are not all necessarily ready and waiting for ministry. In fact, the majority view commitment to their faith in terms of regular attendance at Mass, sending their children to Catholic schools, often at considerable sacrifice to themselves and their standard of living, and general adherence to the doctrinal and moral teaching of the church. Even here, their compliance with church norms is not total. It is clear that vast majority of Catholics of fertile age have made their own decision to use some form of artificial contraception; they also ignore the church's stance on other moral issues, such as some of the teachings on social justice. Most Catholics see their faith in terms of personal and familial commitment and, given the economic and social demands of bringing up a family these days, it is understandable that they do.

Many, especially women, look for a deeper commitment to a wider ministry when their families have grown up. Some Catholics, especially men, see their faith in more subjective terms; for them believing is a private affair and ministry is the work of priests and religious. People with these attitudes are numerically decreasing, as the radically committed group grows, but they are still the majority.

Finally, there are those for whom their Catholic faith is the radical and central focus of their lives and the foundation of their meaning structure. They are still only a small minority of practising Catholics — perhaps only ten to twenty per cent of the group that I have called 'serious Catholics'. However, to ensure its future, the Australian Catholic church must above all harness the energy of these people. In my view, it is these people who will gradually replace the ministerial cadre now provided by religious orders; clearly, some of these lay people will themselves become part of newly emerging forms of religious life. They are people who are already working in Catholic schools as committed lay teachers, or in parishes as pastoral associates. Many of them are in small local groups that are known only to members. Some of the bravest of them are single people, who, in the anonymity of modern life are already living a commitment and ministry worthy of the church's eremitical tradition. Others are divorced people or those who have moved beyond unfulfilling marriages to commit themselves to ministry. The growth of lay theological education is also a sign of increasing lay commitment to the ministry of the church. I will talk about the work of these committed Catholic people in more detail in the sections on parishes, schools and the future of the ministry. The future of the Catholic church is assured, if it is prepared to move beyond the narrow clerical boundaries it has set for itself and to begin to harness the considerable energy within its own lay ranks.

# *Religious orders*

There is no doubt that, especially since the late 19th century, members of religious orders, particularly women, have been the most important ministerial group in the history of Australian Catholicism. [61] There are far more members of religious orders working in ministry today than diocesan priests. But the pattern is changing rapidly. The religious orders have undergone a sharp drop in numbers and, partly as a result, have opted to move out of established institutional ministries, such as schools and hospitals. For instance, Catholic schools in Australia have shifted from having almost ninety per cent religious staff two decades ago to ninety per cent lay staff now.

## The decline of religious orders

The statistical history of Australian religious orders this century is interesting. They increased from quite low numbers at the turn of the century to a high point in 1966 (when there were 19,413 religious sisters, brothers and priests in Australia). The pattern of exceptionally high numbers in the period 1940–1970 was repeated all over the western Catholic world and was, in historical terms, an aberration. All active orders have undergone a sharp decline in numbers since the mid-1960s, with an ever-increasing median age, as few young candidates are joining to begin training. Here are the figures:

---

61. Here I am speaking exclusively about active religious orders. There are contemplative orders in Australia, but the pattern of their life is different and is less subject to cultural patterns than active religious cngregations.

|          | 1901 | 1951  | 1971  | 1976  | 1990  |
|----------|------|-------|-------|-------|-------|
| Sisters  | 3622 | 11245 | 13869 | 12619 | 10074 |
| Brothers | 388  | 1532  | 2221  | 2089  | 1666  |
| Priests  | 195  | 1087  | 2547  | 2321  | 1307  |
| Total    | 4205 | 13864 | 18637 | 17029 | 13047 |

(Source: McCullum, John: 'Secularisation in Australia between 1966 and 1985. A Research Note' in *Australian and New Zealand Journal of Sociology* 23/3(1987), p. 419. Also *Australian Catholic Directory*, 1990).

If you add to this numerical decline an increasing median age, which is now in the late fifties or early sixties, with very few recruits, the picture for religious orders looks very bleak indeed.

Reactionary Catholics are right about the facts: religious orders in Australia (and elsewhere in the developed Western world) are in a crisis of declining numbers and for many orders this crisis will be terminal. But the reactionaries are not right in their interpretation of the facts. The decline is not due to the post-Vatican II abandonment of discipline or doctrine, or because most religious nowadays wear lay clothes. Indeed the fact that some religious orders will probably go out of business is not a reason for panic, for this is neither necessarily disastrous for the church, nor is it a new phenomenon in the church's history. Father Raymond Hostie has shown in *The Life and Death of Religious Orders* that the vast majority of orders founded in the course of church history have already gone out of existence. [62] They have served their purpose.

Here, I think, is the clue to the true purpose of active religious life: 'serving their purpose'. Religious orders were not founded for their own sake and for the sake of the members, but primarily for the purpose of ministry. All their structures should be oriented to ministry and their membership presumably joined to work in ministry. If the ministry is still serving a genuine need, or is still a relevant proclamation of the word of Jesus, then people will join through a natural process. If they do not, it says something about the relevance of the ministry.

---

62. Hostie, Raymond, *The Life and Death of Religious Orders: A Psychosociological Approach*, English trans. by the Centre for Applied Research in the Apostolate, Washington, 1983.

To understand what has happened to religious orders in Australia since Vatican II, it is necessary to review their recent history. Since the mid-1960s, religious orders have passed through a radical renewal in their approach to living a dedicated life and a whole range of plans and strategies have been devised and implemented to achieve this.

One of the most common and overused phrases heard recently in religious orders is 'the charism of the founder'. This refers to the original inspiration and the articulated vision of the person (or persons) who founded the congregation. This 'charism' can be something powerfully tangible if it refers to one of the great religious founders of church history — people of the calibre of Saints Benedict, Dominic, Francis, Clare, Teresa of Avila or Ignatius Loyola. These men and women are examples of members of a group of real geniuses, people of genuine historical significance and originality. Their insight meant that their vision of religious commitment and ministry was able to be raised above the cultural and socio-historical circumstances of their own lives and has endowed the way of life they established with lasting application and value. In other words the Benedictine, Dominican, Franciscan, Carmelite, Jesuit and several other ideals will probably survive current historical circumstances. For they rise above the ruck and their 'charisms' already have a track record.

But the majority of the orders represented in Australia are of 19th century foundation. The visions of many of their founders were often pragmatic and superficial and closely linked to a particular culture, religious expression or specific ministerial work. These visions (or 'charisms') may or may not have continuing meaning. It is not necessarily a tragedy that many of them will slowly wind down or even disappear. The key thing is that their members continue to fulfill their unique role in the contemporary church: that is that they act as educators, facilitators and bridges for the laity to enter into the ministry and replace them. Their final vocation is the most self-giving of all: to hand their ministries over to lay people.

In the process of achieving this some members of religious orders will need to continue to act in leadership roles in traditional ministries such as education, nursing and social service. Others are already moving to new ministries which are genuinely needed but where no one is now operating. Sadly, however, some religious are simply sitting down and wringing their hands, despairing over what is happening. Others have retired to the comfortable existence of living in the false world of an unreal ministry, often associated with some form of 'spirituality'. In fact, one of the saddest things about contemporary religious life is the way in which quite a significant minority of religious hide in ministries which have little engagement with the reality of contemporary Australia. They are not linked to a project involving a genuine educational service, social welfare or social justice, but are usually oriented to individualistic 'fulfillment' or personal enhancement, and almost always involve some vaguely 'spiritual' dimension. They do not liberate or educate the laity to replace religious in ministry. They are the pet schemes of individuals or small groups, who usually have a track record of contributing little to the order which has often provided much for their education.

## Towards renewal

A number of writers, such as New Zealand Marist, Father Gerard Arbuckle, have tried to raise hopes for the continued existence of religious orders by writing about the possibility of 're-founding' them. It is rare that trained church historians write this way, because the evidence from church history seems clear: radical and far-reaching change is achieved by élites within the order who are prepared to confront the structure of a religious congregation and demand that it change radically and quickly. The pace of change is important: if the momentum is lost, the possibility of radical change becomes more and more remote. I believe that many religious orders in Australia have already 'run out of steam', and have lost the opportunity that leads to a new lease of life — which is really what Arbuckle means by 're-founding'.

One of the best analyses of contemporary religious life is by a Canadian, Sister Mary Jo Leddy. [63] She argues that what is happening in religious life today can only be understood from within the context of the wider culture of the Western world. In common with those who say western society is in a state of crisis, she argues that perceptive people in the Western world are experiencing 'an inchoate awareness of being part of a declining culture'. She points out that liberalism has been the sustaining social philosophy of the Western world since the 19th century. However, it is a peculiarly unsuited philosophy for a culture in decline, for it assumes that 'the individual is the starting point in economic, political and social arrangements'. But in societies in decline 'people cease to invest their

---

63. Leddy, Mary Jo, 'Beyond the Liberal Model', *The Way. Supplement 65*, Summer 1989, pp. 40–53.

energy in a common social project and turn to more individually-oriented projects such as personal development and the fulfillment of the self. Evidence of this can be seen in much of the religiosity of the 'New Age' and, sadly, also in some Catholic approaches to spirituality. But, Leddy points out, individualism is unable to respond to 'the very deep human need for common meaning and vision'. The disintegration of 'common meaning' leads directly to the disintegration of the culture.

Reactionary conservatism is also a response to our declining and disintegrating Western culture. Conservatives argue that order and meaning must be restored in society, and that this can be achieved by returning to the traditional moral, theological and social values of the past. But Leddy argues that the conservative vision will also fail, because

a common social vision cannot be imposed. Such a vision arises through the creative, rather than the coercive, use of power. The coercive use of power is a characteristic of an empire in the state of decline. Ultimately, the conservative way of coping with social decline binds its adherents to the extent to which they are subtly perpetuating the patterns of decline even as they attempt to come to grips with its disintegrating effects. [64]

Leddy then applies this analysis of our culture to what has happened in religious orders. She argues that the predominant vision of religious life since Vatican II has been the liberal vision. Not that conservatives do not wield considerable influence, even in quite liberal religious orders. Conservatives try to manufacture coherence and stability by imposing 'guidelines, procedures and structural precisions' in the process of attempting to bring some 'order' to the congregation. Leddy argues that this, ultimately, is a waste of time, for, as she says, order cannot be imposed.

Religious orders came late to the liberal vision of society, but after Vatican II they embraced it with gusto. Leddy says that liberalism has become a most important transitional stage in the development of religious life, but that members must now move beyond it. Congregations, she says, have lost a shared sense of vision and a sustained sense of direction. They have become loose agglomerations of individuals who only have a minimal sense of belonging. 'Liberal communities are held together by an agreement, stated or unstated,

---

64. Leddy, art. cit., p. 43.

to do the minimum together'. But, Leddy argues, individualism sounds the death-knell for the cohesivness of religious orders and ultimately brings about their disintegration.

Leddy says that religious orders have entered a 'dark night', 'when the former models . . . are disintegrating and the future model has yet to become clear'. She emphasises that the dark night will last for some time to come, but also that it is in this place of awkward ambivalence that members of religious orders will not only find their genuine contemporary métier, but also their way into the future. For this to happen two projects must be achieved. Firstly, religious must begin to interact with each other at the deepest level of life, 'where our communion with God coincides with our community with others. She is not just referring to 'shared prayer' here, but rather to a profound communion that is worked at perseveringly over a long period. Secondly, religious need to place themselves consciously at the far periphery of contemporary power structures, where they 'are more likely to feel the need, the hunger and thirst for a different kind of future'. This will lead to a greater pluralism in religious life than exists at present. Out of this matrix alternatives will emerge that will become the basis for the future of religious orders.

This is already beginning to happen. Leddy herself says that she was a member of a small group which founded a newspaper. In Australia there are already groups of religious from different orders working together. Increasingly, men and women religious work together outside the traditional structures of religious life in many ministerial services. Lay people, too, are working more and more with religious; the Catholic school system is an important example. Laity are also involved in the formation of young members of religious orders as spiritual guides and theological educators. There is an interesting example of the way in which this is developing in South Australia. *Adelaide Voices* is a newspaper somewhat like *The Catholic Worker* in approach and format. It is produced by a community of lay people and sisters (mainly of the Mercy congregation), some of whom live together and others who live separately. They work at different jobs — some in the church, some out of it — but they have all made a commitment to the community and to share things in common. It is one of the prototypes of new forms of religious life that are starting to emerge in a number of places in Australia.

## New ways

Finally, as a member of a religious order myself (the largest clerical order in Australia, the Missionaries of the Sacred Heart) I will indulge in an autobiographical reflection on what has happened to religious life in Australia. Thirteen years ago I resigned as parish priest of the order's largest parish, in the Sydney suburb of Randwick, and went into what I then saw as a kind of self-imposed 'exile' in the United States. On and off, I studied and worked in the US for six years. The reason for the resignation, as I saw it then, was the way in which the order was neglecting the parish, especially in terms of the appointment of personnel. To be honest, my action was also motivated by elements of personal pique and considerable anger with the then Provincial Superior. But, although I only came to realise it later, it was also based on a deeper and largely unconscious judgement that the order was simply not facing up to the contemporary ministerial needs of Australia. It was the late 1970s and the heyday in Australian religious orders of 'discerning the charism of the founder' which, in my view, turned orders in on themselves and away from real ministerial needs, emphasising instead the self-engrossed tasks of finding consensus and 'building community'.

I thought then, and I still think now, that the late 1970s and the early 1980s were the lost years, when religious life in this country ran out of steam. For a whole decade the real ministerial action was elsewhere. Fortunately, the orders have, to a certain extent, now turned away from worrying about their own structures and there is a more urgent focus on ministry. But this attempt by religious orders to refocus on ministry may have come too late. A number of religious like me have already had to make all types of accommodations to survive in the active ministry. In fact, my imminent or actual departure from the priesthood and the order has been proclaimed several times by those who thought that they were 'in the know', usually, no doubt, with that indulgent self-righteousness that goes with such *pronunciamentos*.

The key 'accommodation' that I have had to make is to learn how to survive alone, without community support. Certainly, I am an unashamed individualist, who is happiest when working and living on my own. I also need considerable space and independence. But I have had to discover a meaning structure outside the context of the religious community and this has made me realise that a 'shared vision' is a very broad category, and that different people can come to share in it in very different ways. Personally I have experienced the vision of a renewed church in a more eremitical (or hermit) than cenobitical (or communal) way — in other words I have discovered, like Greta

Garbo, that I prefer to live and work alone rather than to interact closely with others in a community.

This is why I think that the views outlined by Leddy on the future of religious life need to be complemented. She apparantly sees no place for the small group of religious who have had to discover a vision of God and a spirituality that supports a life alone. It may be that the lonely life is incompatible with communal religious life as it is presently structured. It may be that I — and a few others like me — are radically wrong. But one of the great graces of genuine Catholicism is that it allows you the liberty to be wrong without casting you into outer darkness! I think that it should also be emphasised that the eremitical life is a strong and lasting tradition within Catholicism. It need not necessarily involve living away from civilization in the wilderness, separate from others, totally oriented to the physically isolated search for God. I see no reason why the hermit existence cannot be lived in the city and strongly oriented to serving God in others.

The essence of all I have said about religious life in Australia is that while many of the current orders will eventually go out of business, they do have the important function of facilitating the movement toward increasing lay ministry. There is also no doubt that some of the current religious orders in Australia will survive and prosper. In fact, many orders not attracting members in Australia, and the western world generally, are doing very well in recruitment in the third world. In Australia new forms of religious life will emerge, some from the current orders, but mainly from the laity. Religious life itself is lay in origin and it is from committed lay Catholics that new forms of religious life will emerge. These new forms will be less structured, more open to temporary commitments, especially among the young, more accepting of men and women living together and not necessarily committed to permanent celibacy; married people will be an integral part of their communities.

In my view, some lay Catholics are already living a radical commitment to Christ, which is what the essence of religious life is. So the answer for religious life in Australia is not the fear-filled, reactionary one of going backwards, but the faithful and trust-filled commitment of moving into a future shaped by God's Spirit. Religious life is an integral part of the church and, like the church, it will survive.

$$\clubsuit\ 3\ \clubsuit$$

# MINISTRIES

## *Parishes and Liturgy*

We know very little about Catholic parishes in Australia, there is very little research available. Given that the parish is the place where most Catholics interact with the church, and where they principally express their faith in worship, this lack of research is surprising. As in so many aspects of its pastoral ministry, the Australian Catholic church has gone along all these years without much thought or reflection on the parish, assuming that all is well in its key pastoral structure. But the question has to be faced: is it?

The Catholic church in Australia is divided into 26 dioceses and the dioceses are divided into 1287 parishes. There are also three eastern rite dioceses (Ukrainian, Melkite, Maronite) with 21 parishes between them. [1] But these statistics tell us little about the nature of parishes that are spread all over the country and are vastly different in size, population and type. Geographically, some rural parishes are bigger than England, others (such as in Sydney's eastern suburbs) have very large populations. The style of ministry called for in far outback Queensland, or a mining town in north central Western Australia, will be very different from the needs of people in an outer Sydney suburb, a wealthy area of Melbourne or a working class area

---

1. *Australian Catholic Directory, 1990–1991.*

in Adelaide, Hobart or Brisbane. Thus generalisations about Australian parishes are difficult to make and usually of little practical use.

It is instructive for our present context to see something of the historical development of the parish. The church in the Roman Empire was almost entirely urban. It only spread to the countryside in the 4th and 5th centuries. In fact, the Latin word *paganus* (pagan) originally meant someone from the country. Churches were only slowly established in the countryside after the Roman Empire became Christian. In the southern part of Roman Gaul, for instance, churches grew up on the sites of old pagan shrines. Parishes developed as the conversion of the countryside in the former territory of the old western Roman Empire proceeded. St Caesarius, Bishop of Arles (c.470–542), was a key figure in the evolution of parishes and it was his canonical legislation that gave some autonomy to local churches. By the middle of the 6th century rural parishes were spreading throughout western Europe and the network became thicker in the succeeding centuries, reaching England in the 8th century and Germany just a bit before that. The parish has always had a priest; in fact, in early medieval times the parish actually developed as a unit sufficient to support one priest. In the Middle Ages the church became almost entirely oriented to a rural ministry.

In the following centuries, radical changes in social structures occurred with the growing urbanisation of most of Europe and then the rest of the world, reaching its peak in the contemporary period. But our ecclesiastical structures are still modelled on those that developed in a fundamentally European rural world. It has often been pointed out by recent writers that the parish is a static and non-missionary structure, reflecting an established social order and it provides a comfortable sub-culture for many people.

The link between priest and parish was also strong from the early Middle Ages onward. The parish priest was a recognised part of the social fabric of European culture. Today, however, we may be seeing the breakdown and even the end of this long-term connection between parish and parish priest. The implications of this shift will be far-reaching for both parishes and for priests.

Even in a country like Australia, there are fewer and fewer priests to go around and the median age of the clergy is rising rapidly. A clear example of this emerges from the work of the Melbourne Catholic Research Office for Pastoral Planning, one of the few sources of contemporary data on priests and parishes. In 1977 there were

340 diocesan priests in Melbourne parishes; in 1990 there were 289; in the year 2000 there will be 245. In 1977 the average age of the active diocesan clergy was forty-four; it is now fifty-two and in 2000 it will be fifty-eight. [2] At present the Melbourne Archdiocesan Catholic population of probably close to one million is served in parishes by 289 diocesan priests and about 35 religious order priests. The Melbourne average is about 3000 Catholics to one parish priest. [3] The national average is lower: about one priest to 2000 Catholics. If we presume that one third of these people are more or less practising Catholics, it means that the approximate national average is one parish priest serving every 670 active Catholics.

There will, of course, be many more people in urban parishes than rural ones, and this shows up in the average figures. The diocese of Parramatta, in Sydney's western suburbs, for instance, averages one parish priest to 2400 Catholics, but the diocese of Bunbury, in the south west corner of Western Australia, has one parish priest to 1200 Catholics. The diocese of Townsville, taking in much of the central coastal and inland area of Queensland, has one priest to 1500 Catholics. [4] But a priest in a rural parish in Townsville diocese has to travel vast distances to serve his parish, while a priest in Brisbane or Perth or Canberra or Hobart has the suburban congregation pretty much on his doorstep.

Further, priests are faced with the fact that their responsibility is not just to the *active* Catholics; they are there to minister to all who are baptised members of the church. So they cannot just ignore the non-practising, or the occasionally-practising Catholics — who do appear regularly enough for the sacramental celebrations of life transitions. However, if priests are already working hard at the immediate ministerial needs of the active Catholics — and a lot of priests are — then there is not much they can do for unchurched people.

It is sometimes pointed out that the average ratio of Catholic priests to people is in sharp contrast to the ratio of clergy to people in the other Christian churches. Taking the figures provided by *A Yearbook for Australian Churches 1991*, the Anglican diocese of Adelaide,

---

2. *Cropp News*, 1/2(August 1990), p. 3.

3. Here I am using the term 'parish priest' not in its technical canonical sense of the pastor in charge, but to refer to priests actually working in parishes as distinct from other ministerial works.

4. All these figures are based on the absolute number of Catholics in the diocesan area; they do not refer to active Catholics, who can be assumed to be about one third of the total numbers.

for instance, has one priest to every 165 attenders, Newcastle has one to 370, Melbourne one to 170 and Bendigo one to 100. The Protestant figures are even more startling: Victorian Baptists have one pastor to every 90 active members. The Lutheran national average is one pastor to every 40 baptised members. [5] Similar comparisons can also be found in the United States, where it has been demonstrated that in some places the population of the average Catholic parish is twelve times that of the average Protestant parish.

These comparisons may be startling, but they can also be misleading. Catholics, by and large, do not expect as much of their clergy, in terms of social contact and pastoral care, as do Anglicans and Protestants. Catholicism is better adapted, it seems to me, to the anonymity of urban life. While belonging to the parish is important to many Catholics, the primary expectation of most of them is that the church provide sacramental services. In the past, the confessional was the one point of intimate pastoral contact between Catholic priest and people. With the effective breakdown of this sacrament at present, many Catholics have learned to be increasingly spiritually self-sufficient.

Thus it is difficult, if not impossible, to decide on an optimum ratio of priests to Catholic people. It all depends on what the parish expects of its clergy. If the parishioners want the priest to fulfill the entire institutional, spiritual, liturgical and administrative gamut of church ministry, then the non-Catholic churches are probably closer to the required ratio. But if the priestly task is one primarily of leadership, especially in worship, with lay people increasingly carrying out other ministerial tasks, then the Australian Catholic church needs more priests than it has now, but not necessarily a lot more. Much, too, depends on the quality and ability of the clergy that the church has. As I have already said, the spiritual and psychological health of many of the clergy leaves much to be desired. It is to be doubted whether they can take on much more.

From one perspective, the decline in the number of Catholic priests is a golden opportunity for the many women and men in the church who want to participate in ministry, but who have been excluded by the clerical monopoly of so much of the church's work. It is ironic that a church which has been so often accused of being 'priest ridden' is the one that is now short of clergy. If the present shortage of priests continues — and there is no reason to expect that it will change,

---

5. *A Yearbook of the Christian Churches 1991,* pp. 145, 146, 148.

especially in the pontificate of John Paul II — it is clear that the Catholic laity will be thrown back increasingly on its own spiritual resources. The shortage of priests creates the ministerial space where gifted and trained lay people can begin to exercise all types of service.

## Eucharistic celebration

However, the priest shortage has created one major problem. Australian Catholics are usually realistic and do not expect their clergy to be great preachers or spiritual gurus, but there is one thing that they do expect: that priests will provide them with the sacraments, especially the eucharist. The shortage of clergy means that even now in parts of Australia there are parishes without priests, resulting in what Father Bernard Lee of the Institute for Ministry at Loyola University in New Orleans, calls 'eucharist-less' parishes. [6] On a world-wide scale, the shortage of priests means that an ever-increasing number of Catholics are deprived of the eucharist for long periods.

This has already been a problem in Latin America for two decades; it now effects parts of Asia, Europe and increasingly North America. In Indonesia, for instance, the clergy shortage is acute. This is vividly illustrated in Irian Jaya. The Catholic church has grown rapidly there since the 1950s. Indonesian Carmelite Bishop, Franciscus Hadisumarta of Sorong, Irian Jaya puts the problem clearly:

The eucharistic celebration is said to be the peak of a Catholic's liturgical life, but here Catholics celebrate only the liturgy of the word on Sundays. Even on Christmas or Easter, it is very common that Catholic communities won't celebrate the eucharist. Rome suggests that the liturgy of the word can be followed by holy communion. But that is impossible in many places because we cannot consecrate hosts sufficient to last several months. [7]

The Indonesian bishops have literally pestered Rome for permission to ordain suitable married men, but have been consistently refused.

This is a totally unacceptable situation. It means that the Catholic church is being increasingly 'Protestantised', in the sense that many Catholics can only gather to hear the word of God, but not to celebrate the eucharist. This is completely at variance with the church's entire historical experience. In fact, to deprive baptised and confirmed believers of the central sacrament of the church is so untraditional that it amounts to practical heresy. Heresy is not just

---

6. See *National Catholic Reporter*, 13 April 1990.
7. Quoted in *National Catholic Reporter*, 22 April 1990.

wrong belief; it is also wrong practice. The blame for this situation lies squarely with some of the bishops, the Roman Curia and Pope John Paul II. In this situation of eucharistic deprivation, the real radicals are not those who suggest that the church ought to bring back as celebrants priests who have left the ministry to marry; nor those who say that non-celibate and suitably trained men ought be ordained; nor even those who say that women ought to be ordained as deacons and priests. The outrageous radicals are those who say that Catholics should be deprived of the eucharist in order to maintain the clerical custom of celibacy, the male dominance of the priesthood and total Roman control over the local churches, who, in any case, ought to be able to make up their own minds on this central pastoral question.

## Leading the community

There is another area besides the celebration of the eucharist where the priest has a real and vital role and whose importance will increase in the church of the future. It is the leadership of the parish community. Every parish needs someone who acts as a leader, someone who symbolises and represents within themselves what the whole community is about. In certain élite groups, such as among some religious orders, this function may be able to be fulfilled by a group or team, in a totally non-hierarchical way, but I do not think that this can happen in the ordinary local church community. In the parish someone must stand, not necessarily *in persona Christi* ('in the person of Christ'), which is a function that will probably be fulfilled by the whole community, but at least *in sancto spiritu* — 'in the Holy Spirit'. This is the person who will give a sense of unity and direction to the community, who will discern and animate the gifts of others and draw all the threads together. Increasingly in Australia, the best priests are already assuming this difficult and self-effacing task. In the United States, the priority in priestly in-service training has been to emphasise the development of appropriate skills so that priests may begin to assume a non-directive form of leadership in their parishes.

Bishop George Pell seems to take account of none of this in his rather quaint vision of the future church in the year 2000. Since he is emerging as a spokesman for some elements in the Australian church and has also been appointed to two Vatican bodies, as well as to a major position in the new Australian Catholic University, anyone interested in the church needs to examine seriously his attempted exercise in ecclesiastical 'futurology'. What he describes in the magazine *AD 2000* is, in fact, a rather unimaginative scenario that harks back to the church of the 1950s. [8]

After assurances that the parish will remain 'the basis for most of the church's pastoral work' and that the Mass will remain central ('but celebrated only by ordained priests'), he then goes on to say that 'individual confession and reconciliation will be stressed . . . and the celebration of the third rite of penance will be only a memory'. Here, of course, he is referring to that bête noir of the reactionaries, general absolution without individual confession of sins to a priest. He then assures readers that 'devotion to Our Lady will have experienced a resurgence . . . [and] devotion to the saints will be revived'. The idea of the ordination of women will only be a bad memory kept alive by 'a small number of people'; the need for communion services without a priest will have disappeared because there 'will be a rise in priestly vocations'. The colourful picture continues:

There will be no priestless parishes in the urban areas, although a number of parishes will have been combined, and the number of priests in specialist positions and as curates had been reduced. The religious leadership of priests in all parishes was a constant, although this leadership was always exercised with parish council and lay workers [sic].

The reference to 'priests in specialist positions' is significant: for earlier in his futuristic scenario, Pell makes it clear that he shares, with most reactionary Catholics, a particular distaste for church bureaucracies, especially when they are under the control of 'liberal' clergy and laity.

For Pell's vision to work, the pivotal issue will be the need to increase the number of clergy. He has a ready answer for this problem:

Vocations to the diocesan priesthood also increased [in the 1990s]. All of this was helped by the catechetical renewal which followed on the introduction of the Universal Catechism in the early nineties.

Mention of the *Universal Catechism* implies that there is a direct correlation between 'right doctrine' (which, of course, can only possibly come from a document like a Vatican catechism) and an increase in the number of vocations to the priesthood. Incidently,

---

8.   Pell, George, 'The Future of the Local Parish', *AD 2000* 2/9(1989), pp. 10–11.

the *Universal Catechism* has not yet been published; it is not expected for a couple of years. There is no evidence for any causal connection between 'right doctrine' and vocations to the priesthood. I would presume, for example, that 'right doctrine' reigns supreme in Melbourne, the diocese in which Pell is an auxiliary bishop. But there were only twenty-five seminarians for the Melbourne archdiocese in 1990. The Melbourne Catholic Research Office for Pastoral Planning says that these twenty-five seminarians 'represent, at best, an average of 3.6 ordinations for the next seven years, but the departure rate from the seminary over the last thirteen years has been 56%'. [9] That same picture can be reduplicated in seminaries right around Australia. The problem is not only the question of the number of seminarians and ordinations to replace the present clergy; it is also a question of the high average age of the clergy. I do not know what evidence there is for Pell's optimism about an increase in vocations to the priesthood, when the evidence of his own diocese's research office is against him. Not fazed by statistics, Pell is reported to have said: 'Those who predict that the present dearth of vocations will continue are wrong'.

Pell returned to the theme of the centrality of the priesthood in the future ministry of church in an address he gave recently. [10] He argued that there was a danger in discussing options for future lay ministry because it could unconsciously undermine the morale of priests. Statistics about the shortage of clergy were a 'beat-up'; the ratio of priests to people in Melbourne was very favourable when compared to statistics in South America and the Philippines. He is certainly correct that the Australian ratio of priests to people is higher than Latin America or the the Philippines. He forgets that the situation of the church is very different in both those areas to Australia and that it is hardly a flattering comparison, given the acute shortage of clergy in these Latin countries.

Everything in Bishop Pell's argument comes back to two predictions. Firstly, that there will be an increase in vocations to the priesthood. Secondly, that the model of the church operative from

---

9. *Cropp News* 1/2(1990), p. 5.
10. Bishop Pell was addressing a meeting at Saint Peter's church, East Bentleigh on 14 December 1990. The meeting was called to assess the Catholic Research Office for Pastoral Planning's discussion paper 'Pastoral Planning for the Future of Parish Leadership'. It was at this meeting that he said that the planners were 'wrong'.

the Council of Trent (in the 16th century) to the period after Vatican I (1870) has not changed since Vatican II. Pell seems to argue, in fact, that the model of church has not changed at all since the period of the apostles. This is a remarkably static view of such a dynamic reality as a church which officially teaches that the development of doctrine is part of faith. It is doubly unusual, since Pell is a historian by training.

As an exercise in 'futurology', Pell's scenario also seems to operate from the presupposition that Australian Catholics will not have changed their religious and spiritual priorities by the year 2000; that the attitudes of 'the Catholic in the pew' will be what they were twenty years ago. There is no evidence for this. He makes much of the multicultural nature of the church and emphasises that foreign-born Catholics will want to hark back to the pre-conciliar model of the church of the 1950s. As I have already pointed out, the religious priorities of foreign-born Catholics are more subtle than the mere desire for a restoration of past. Bishop Pell has described himself as a 'romantic conservative'. While it is clear that both words are operative in this self-definition, it is obvious from his paradoxically reactionary vision of the future, that the stress is on the word 'romantic'.

## The Sunday liturgy

To return to the present reality of parish ministry, let us look at the place where most ordinary Catholics encounter the church in action every week: Sunday Mass. The sad reality is that most liturgy in Australia is dull, lifeless and boring, and in some places it is quite appalling. You cannot blame people for not going; the miracle is the faith of those who still do. Given the sexist attitude and language manifested by so much Catholic worship in Australia, women in particular require a special commitment to keep going.

Obviously, there are quite a few places where this sexist lifelessness is not found and there is good worship. But the majority of parishes in this country are liturgical wastelands. Most Catholics blame the clergy for this. The low standard of preaching is often cited, as well as the failure of many priests to be effective leaders of public prayer. But clergy are not entirely to blame. The ability, or lack of it, of particular priests as celebrants is certainly part of the problem, but it is clear that the core of the liturgical problem is really socio-theological.

One of the most marvellous liturgical experiences I have ever had was in the tiny central Pacific nation of Kiribati. In the cathedral at Tarawa the celebration of Sunday Mass literally 'took off'. There were no gimmicks, the Roman Missal was followed to the letter, and the potency of this eucharistic celebration had little to do with the celebrant. The source of the power of the experience was the worshipping community itself. At the external level this was achieved by the marvellous choral singing in which everyone joined there was a real sense of participation. Considerable ability in choral singing is a characteristic of most Pacific people, but the celebration in Tarawa cathedral went much further than this.

At the deepest level the liturgy was good because it was the expression of a genuine community, not in the sense that everyone knew each other (although they probably did), but because the participants shared a common culture, a common attitude to life and a common faith that had transformed them, both as individuals and as a community. In other words, their culture generated a sufficently shared experience of faith for their worship to release among them a deep sense of entering together into the presence of transcendent reality.

In Australia, Catholics simply do not have a deep enough common cultural experience upon which to base a faith that can sustain a coherent and living liturgical celebration every Sunday. It is not that faith is lacking. Australian Catholics believe as deeply as Pacific people, or anyone else in the world. Certainly part of the problem is that Australian culture emphasises an individualism that defines people in terms of their differences, rather than what they have in common. Another part is that Australian Catholics come from many different cultures and their approach to life generally, and to religion specifically, is very diverse. Australia has made the policy of multiculturalism an important value in national life. We treasure our individual and cultural differences; what we have in common as Australians is often forgotten.

This combination of individualism and multiculturalism creates real difficulties in achieving that common culture, spirituality and religiosity that gives a congregation a sense of liturgical coherence and thus an opportunity to move together into the presence of God. I am not saying that this cannot be achieved. The problem is that the lack of a common culture is insufficiently recognised. As a result no coherent attempts are made to circumvent or bridge the cultural and spiritual gap between the individuals that make up Australian congregations.

Not that the Australian Catholic church fails every time: a feast like Christmas can be an expression of enough common cultural ties to bring people together, in a worship that creates enough communal sense for them to share in an experience that often deeply affects them. Of great significance in this is the common musical culture that is represented by the carols.

Australian Catholics might experience better worship if Mass was celebrated in a way that released and channelled the profound and common elements of Catholic faith that is shared by all those attending. But Catholic worship in Australia is often perfunctory, unprepared and unimaginative. The music, which can provide a common element that draws people together, is almost universally bad. The result is that the congregation feels unmotivated and bored.

Let us stay with the theological theme for a moment. It is time for reformers and 'liberals' to admit that in an attempt to give people greater access through liturgical reform the church since Vatican II has, in fact, lost a key defining element of genuine Catholic worship: that is that worship is a form of sacramental symbolisation. By this I mean that the celebration of a liturgical action should shift those participating out of everyday reality and move them into a sacramental space which transcends time and within which the risen Christ is personally present. It is in that sacred space that the salvific power of God is released for those participating.

Genuine worship is an evocative action that resembles the effect that great art has upon us. But for this effect to occur the external circumstances have to be right. There has to be an outer liturgical form that has shape, discipline and beauty, that will attract those participating in the worship to the journey inward where they can discover and experience the living presence of God. Sadly, in most Australian Catholic churches, Mass is just a text that is read badly and a ritual that is gone through, often by an undisciplined and garrulous clergyman.

There is nothing new, mind, about bad liturgy in the Catholic church, both in Australia and elsewhere. Bad worship is not something that has been invented since Vatican II. Many Catholics, especially the younger ones who never knew the old Latin liturgy, romanticise the past and pretend that the 'old' Mass was somehow perfect and that it was always celebrated with great faith and with profound spiritual and mystical effect. This is simply not true, for it ignores the human factor of the limitations of the priestly celebrants. Having been an altar boy for several years I can testify that some pre-Vatican

II celebrants distinctly lacked any real spiritual insight or style. As in so many other things in the contemporary church, the answer is not to attempt to resurrect the liturgy of the past, but to move creatively beyond the present.

Recently, I attended a concert of late medieval and renaissance music sung in Saint Patrick's Cathedral by the Tudor Choristers. For me it was a transforming religious and artistic experience. I have a special affinity to Saint Patrick's, having served on the altar there in the early 1950s, in the latter years of Archbishop Mannix's episcopate, (as did my father and uncle before me). It is the finest gothic church in the country and probably one of the best of the churches built in the neo-gothic style in recent centuries. So I was well prepared through my affinity with the church for a transforming experience. But beyond this special connection, the music, the church, the sound and the ambience all merged into each other in an extraordinary way. It was an expression of the fact that all were designed to go together. Of course, it is true that the worship of the church is not a performance in the sense that this was, but there has to be an element of good drama in any genuine worship. When a believing community and well-performed liturgical action are merged into the matrix of fine sacred space, accompanied by well-composed music, you can have transformative worship. Certainly, the ordinary parish church usually has none of the ambience and professionalism of a cathedral working for it, but this does not excuse the poverty of so much ordinary worship in Australian parishes.

## Achieving commonality

However there are a number of positive aspects working even in the most difficult parishes. There is usually a believing congregation present. Most people who go to Mass these days do not go because they have to, because it is a mortal sin to miss; they go because that is what they want to do. I am not one of those who think that you will improve liturgy by gimmicks or by turning it inside out. Nor do I think that the church has to develop a so-called 'sense of community', as though this could be manufactured by some sort of social engineering. I think that much contemporary worship has often been ruined by those who know nothing about how strangers and disparate people interact with each other in a large social situation. In heavy-handed ways they demand that people 'relate' by holding hands or introducing themselves and exchanging pleasantries.

The public worship celebrated in most Australian parishes does not need such artificial 'community' but rather a sufficient sense of commonality for the participants to share a common, but often unspoken sense of purpose, which is an entry together into the transformative presence of God. Genuine community is characterised by interactive, personal and supportive relationships. It is obvious that this cannot and, in fact, probably should not be what parish worship is all about. Community is a voluntary association, achievable only in a small group. Parish worship is characterised by a number of different people coming together for a common purpose, which is both the worship of God and a sense of communion with the other members of the Catholic community. The idea of commonality is something very public, while community is basically a private phenomenon. As Niall Brennan has pointed out, Mass is one of the few places in Australian life where class and vocational distinctions are forgotten, where the rich and the poor, the powerful and the weak, stand together on an equal footing before God.

The sense of commonality is released and channelled by the creation of the right type of atmosphere. It is not necessarily difficult to achieve this. The place of celebration is an essential part of the sense of commonality. There has to be a feeling of stepping outside banal and everyday reality to enter into God's presence. Homes and classrooms might be the right places for prayer; they are often not the right places for liturgical worship, which requires a sense of entry into the numinous. There are a number of fine parish churches in Australia, both traditional and modern. Some are very well adapted to the renewed liturgy. Although many other churches are not artistic masterpieces, they are workable for worship and could be improved interiorly with knowledge and taste and by the shifting of some of the furniture. For genuine worship people need sacred space, a place with some distance from common reality, which releases within participants the power to contemplate a Reality that transcends and expands our understanding and ordinary human experience. That is why I think that Mass should only be celebrated in a church or in one of those marvellous sacred spaces of which the Australian bush is full.

Another key element in the achievement of commonality is music. Much of the hymnody used in the contemporary Australian church is tuneless, difficult to sing and musically cliche-ridden. It is antithetical to good worship. In the last twenty-five years there have been great efforts made to improve the theological content of hymns by making the lyrics more scriptural. This is based on the assumption that the words of hymns are vitally important and that the theology of the hymnody of the Victorian era, which formed the staple of singing in Catholic churches prior to Vatican II, is unsuitable for modern congregations. I sometimes suspect that Father Faber and his Victorian ilk knew a little more about church singing than the Saint Louis Jesuits and most contemporary liturgical composers! [11] However, both Faber and modern hymn writers have one thing in common: most of their hymns are so bad that they have been, or hopefully soon will be relegated to the oblivion they so richly deserve.

Where Faber and the hymn writers of the Victorian era were right — and contemporary hymn writers plainly wrong — is that they understood that the melody of the hymn is much more important than the words. Modern hymn writers have clearly shown that good theology is often very bad music. Many regularly performed and melodically enjoyable operas have quite ridiculous plots and silly librettos. Yet people keep going to these operas because they love the music. It is the music, the melody, the emotion evoked by the tune that draws people together and gives them a sense of commonality.

In saying this I certainly do not want to exclude good contemporary music and I am not recommending a return to a diet of 'Hail Queen of Heaven' and 'Faith of Our Fathers', although both of these hymns have fine melodies! But I think that parishes have a right to demand of those who lead their singing that the banal, the sentimental and the musically inept and ugly, no matter how modern or theologically correct, be excluded from the liturgical repertoire. They also have a right to demand that the melodies be both tuneful and singable, that they retain their emotional and artistic integrity by being words set to music, not music set to words, and that they be accessible enough for the majority of those participating in the liturgy be to drawn into the service, rather than alienated.

There is a need in hymnody for a renewal of taste, for a sense of what is right and proper in the worship of God. The fact that someone *likes* a hymn does not mean that it is tasteful, or that it should be imposed on long-suffering congregations who have been drowned for so long in bad taste and sentimentality posing as theology

---

11. Father Frederick William Faber was the writer of many Victorian hymns such as 'Faith of our Fathers', 'Mother of Mercy' and 'Sweet Sacrament Divine'. See Ronald Chapman's *Father Faber*. London, Burns and Oates, 1961.

that they have lost their ability to distinguish the good from the bad. So often hymns are imposed on the congregation by a small group of people who usually have little notion of what the renewed liturgy is all about, many of whom have no training in music. They have attained their position of influence by default, because the Australian church simply has not taken church music seriously. There has been a total lack of leadership shown by bishops and priests in the area of musical taste. The fact is that next to none of them knows anything about liturgy, let alone church music. Yet music is the most potent influence that can be used to draw a group of people together and help them attain a sense of commonality sufficient for worship.

It was not only Father Faber and the Victorian hymn writers who were literally cast out of the church after Vatican II to be replaced by 'folk hymns' — some of which are far more sentimental and indecently narcissistic than anything Faber or his contempories ever wrote. The tradition of Latin plainchant, classical polyphony (which includes composers of the calibre of Dufay, Josquin, Palestrina, Victoria, Byrd, Lassus and Monteverdi, to mention just a few composers from the 15th and 16th centuries) are unknown to most Australian Catholics, unless they are regular listeners to serious music stations or attend Mass in those few Australian cathedrals or city churches that present such music. A whole generation of young Catholics has lost touch with this musical culture. In those Catholic secondary schools where there is an attempt to create an awareness of this culture students are surprised, often very pleasantly, that they are the inheritors of such a wealth of tradition. The traditional musical culture is an inherent part of the Catholic belief system; it is important that it be renewed.

The development of a 'sense of communion' is another way of overcoming the obstacles of individualism and the lack of a common culture. By sense of communion I do not mean that everyone has to know everyone else personally, or that everyone be friends. I, for one, cannot tolerate instant attempts to develop intimacy with too many people. Most Catholics do not go to Mass to overcome their social alienation, but primarily because they want to enter into the presence of the living God. This word 'presence' is a key to understanding the genuine communion that draws a congregation together at the deepest level. Impelled by faith, the members of the congregation share in a common desire for the infinite and, in optimum circumstances, they are peculiarly and uniquely open to God and to each other. Ham-fisted attempts at intimacy, like bad music or unsubtle and talkative celebrants, simply impede or destroy this openness. The more the focus is on creating an awareness of God's presence, the more a deep and pervasive sense of communion is generated within the worshipping congregation.

I have compared good worship to art. Understanding the way in which art works is a genuine key to understanding how liturgy works. Susanne Langer points out that art exists in real time and space, but it has the facility to transmute the perceptive viewer or listener out of ordinary time and space and to draw them into a transcendent dimension that is fundamentally numinous or holy. [12] That is why the space and ambience created by word, action and music must be oriented to taking the congregation out of the everyday to enter into a presence that is transcendent. Paradoxically, this very movement away from the banal assists participants to enter ever more deeply into the meaning structure of their everyday lives.

In everyday life we exist in one dimension of time which is chronological. But good worship, like great art, moves us into the depths of time. It happens in time, but it places us in contact with a reality that transcends time. It lifts us out of ourselves, evokes depths within us that we rarely experience. At their best both art and worship move participants into that intuitive form of knowing which is called 'perception'. By this I mean that we move beyond the hubbub of habit and everyday life into a realm in which we can 'see' more deeply.

Genuine worship also demands that the congregation participate and be part of the faith action. For this to happen the congregation cannot merely be passive spectators viewing something going on 'out there', but participants in the fullest sense. I do not mean this in the crude sense of actually doing anything external. I am referring more to the subtle sense that pervades a congregation that they may not even be aware of themselves. A celebrant can actually experience this type of 'living' congregation. I have been touched by this vividly on occasions. There are times and places where the congregation is noticeably alive and, as celebrant, you can experience a quiet, participative faith which is sustaining and supportive in the very process of saying Mass. At other times, the congregation can feel like a dead weight that weighs you down and is unresponsive no matter how hard you try to lift their spirits.

---

12. Langer, Susanne, *Philosophy in a New Key*, Harvard University Press, Cambridge, Mass., 1951.

## Sacrament vs word

I think that one of the major reasons why our worship is not working is surprisingly a result of the biblical renewal that has been going on in Catholicism since well before Vatican II and which has blossomed since. Because of the biblical renewal, there has been a subtle shift of emphasis away from the importance of symbol and sacrament towards the word of God; away from worship as a symbolic action which renders Christ present, to the reading and interpretation of the bible. The major reason why I think this has happened is the orientation of post-modern western culture toward the superficial accessibility of words and away from other deeper forms of symbolic understanding. In this refocusing on the word of God Catholics have moved toward the Protestant end of the Christian spectrum and in doing so are in danger of lessening their sense of sacramentality.

One of the great tragedies of the 16th century Reformation was the loss by the Protestant tradition of a sense of sacramentality and the importance of symbolism. The insistence on *sola scriptura* — the bible alone — meant that Protestant theology became dominated by the word of God to the detriment of other, sometimes more powerful symbols of God's presence. Thus the sermon and the reading of scripture became all-important in Protestant worship. Perhaps it was only in their hymnody that Protestants retained a gateway into a less tangible, more subtle form of God's presence. Paradoxically, at the very time that Protestants are rediscovering the sacraments and the liturgy (in Australia this is illustrated well by the renewed worship of the Uniting church), the Catholic church, in the process of rediscovering the bible, has become overly dominated by the word. Theodore Roszak's comment that mainstream Christianity 'has sacrificed the rhapsodic for the word' is, in my view, both entirely accurate, and also very dangerous. [13]

Why is Christianity so dominated by the 'word'? While Christian theology, especially in the West, has long been dominated by the importance of words and dogmatic statements, historically, domination by the word has only been an issue outside theological circles since the 16th century. It is clearly related to the development of printing and the increasing growth of literacy. While I am not recommending a return to illiteracy, I think we have to recognise

---

13. Roszak, Theodore, *Where the Wasteland Ends: Politics and Transcendence in Post-industrial Society*, Anchor Books, New York, 1973, pp. 349–378. The whole chapter 'Rhapsodic Intellect' is well worth reading.

one of the consequences of literacy: it can debilitate the symbolic and intuitive functions of the intellect and it exalts the rational at the expense of the visual.

Indeed, in spite of the spread of literacy, many people in our culture are still functionally illiterate. This is why television is such a powerful cultural medium; it communicates fundamentally through visual images rather than through concepts. It is a difficult medium to use for the discussion of abstract ideas; it is marvellous however for communicating powerful visual images. It has been such a Godsend for the environmental movement, for instance. You can describe in words the slaughter of the seals in Labrador, or oil pollution in the Gulf, or the destruction of the rainforests. But one television image is worth twenty thousand words: you can see the dying seals, or the result of Saddam Hussein's environmental vandalism, or the timber companies raping the forest. That is far more powerful than all the talk in the world about 'sustainable development'.

Modern Australian Catholics are going to have to re-discover the symbolic and the visual if the liturgy is going to be rescued from its contemporary debilitation. The symbolic way of thinking is deeply embedded in the Catholic tradition. I once spent a fascinating morning in Chartres cathedral exploring the iconography with the brilliantly eccentric English guide, Malcolm Miller. I learned more that morning at Chartres about symbolism and the way that ordinary, pre-literate medieval men and women thought than I had in all my history courses. The 12th century French cathedral is full of iconography because medieval people thought symbolically. Everything in the cathedral — the stained glass, statues, gargoyles, even the very shape of the architecture spoke of the transcendent to the perceptive medieval woman or man. They thought in images that opened for them a world filled with meaning.

We have lost virtually all of our ability to think in this way — though not all of us. The medieval way of thinking is analogous to the way that Aboriginal people think, though for the Aborigines it is not a humanly constructed cathedral but the land itself that is full of the symbols that conjure up the cosmic drama in which transcendent meanings are constantly played out. For the Aborigines, Australia itself is a vast cathedral. In order to participate more fully in the liturgy, Catholics will have to re-discover that in addition to the language of words, there is the perceptual language of the senses and the heart. The renewal of the worship of the Australian Catholic church will not be an easy task.

## Preaching

One of the aspects of liturgy most criticised by Australian Catholics is preaching. Local Catholic clergy are probably no worse in this regard than priests in other English-speaking countries. Some of the worst sermons I have heard were in the United States, where I once heard a priest at a marriage ceremony preaching against abortion. He may have felt righteous, but he certainly lacked a sense of occasion. The Australian record on preaching is not good. Probably the only significant Australian preacher of this generation is the Reverend Sir Alan Walker of the Uniting Church. Certainly, there are many individual clergy who have helped people through their preaching and said the right thing at the right time. Ironically, it is my own experience that it is on the those occasions when you feel you have made a fool of yourself in the pulpit that someone will come up after Mass and say that what you said helped them enormously. It is clearly part of the Holy Spirit's odd sense of humour!

To a certain extent contemporary preachers are faced with an almost impossible task. They begin behind scratch because there is an unconscious assumption in our culture that preaching is boring. As if to re-enforce this, there is a close connection in popular rhetoric between those exact words, 'preaching' and 'boring'. In addition, preaching is difficult: it demands that the priest be able to interpret a particular text of the bible in the light of contemporary experience. This requires a double-edged skill: a good exegetical knowledge of the biblical text, which some clergy have, linked with a maturity of faith and a broad experience of life, which is much more difficult for any human being to obtain. Priests, especially those with immature faith, are often very exposed personally when preaching, far more than most of them realise. I remember once talking to a very perceptive friend some years back about a certain priest. She said that when she listened to his preaching she wondered if he believed in what he was saying; in fact, she wondered if he believed in anything at all. He left both the priesthood and the Catholic church a few years after she made the comment.

While the preacher has to face all these difficulties, he (and it is almost universally 'he') has one enormous advantage working for him. There is a real hunger in the Christian community for insights into the biblical word of God and into the Christian interpretation of the meaning of human existence. People keep coming back again and again to Sunday Mass hoping for some insight into a deeper faith and spirituality. This is a tremendous opportunity to exercise the church's ministry.

One of the major problems with preaching is that it provides an almost irresistable opportunity for clerical self-indulgence. For example, there are those preachers who return again and again to their pet biblical, spiritual or theological topics. One of the great disciplines that result from the use of the lectionary at Mass is that the readings of the day should determine the focal topic of the sermon. But that does not stop some! I will never forget a sermon I heard in a well-known church in Australia in which the preacher twisted the biblical text around so that he could denounce the state of religious education in Catholic schools. This then led him to give a discourse on the fact that Catholic students were also poor at 'mental arithmetic'! He had obviously not yet caught up with the invention of computers. The adults in congregation simply looked dazed, the young people looked bored and I was seething! I felt that his manner was brutal and harsh, although friends at the Mass saw the humourous side and said he reminded them of the English comedian, John Cleese.

Other preachers project their unresolved conflicts by constantly telling the long-suffering congregation about their personal weaknesses (such as their alcholism), or, even worse, by projecting their conflicts outward onto the congregation and accusing them of problems that are clearly personal to the preacher. Some priests also project their immature moralism and their personal fixations. I will never forget a preacher I heard shouting at a congregation 'You are commanded to love' as though it was an order from a regimental sergeant major. Jesus' own command ('Love one another as I have loved you') is rather an invitation than an order. Lucinda Vardy tells the story of a callow young priest in the cathedral in Toronto, Canada, who used one of the incidents of casting out of a devil in the early chapters of Saint Mark's gospel, to lecture the congregation of middle-aged, well-heeled adults on masturbation, pre-marital sex, birth control and abortion. [14] I have always thought that pastorally inexperienced priests should not be allowed to speak publicly about sex, and that this topic should only be dealt with in the pulpit with considerable circumspection. There is a danger that immature preachers of whatever age can play on the vulnerability of psychologically and spiritually fragile people in the congregation.

---

14. Vardy, Lucinda, *Belonging: A Questioning Catholic Comes to Terms with the Church*, Macmillan, South Melbourne, 1988, pp. 134ff.

And while such priests can do damage to the vulnerable, they simply drive to distraction those who are spiritually and humanly mature.

These horror stories about preaching can go on and on, but a more important question is: what is it that Australian Catholics want from a sermon? I believe they do not want great eloquence, just a feeling that the preacher has some appreciation of the reality of their human experience. Having established some empathy with the congregation, the preacher then has to examine human experience and reality in the light of biblical and church tradition. In this process he must challenge both the congregation and himself to move just a little further forward in spiritual and human growth.

One of the most important aspects of preaching is the prophetic dimension. There is a strong biblical and Christian tradition of the need to confront the self-interest, blindness and selective amnesia that is deeply ingrained in all of us. But these destructive realities are not just personal; they are also structural. They are built into the very fabric of our society. Religion has often been used and still is used to support exploitative and totally unjust social structures. For example, we belong to a culture that is so self-engrossed that it is prepared to use up all available resources, leaving nothing for our grandchildren and destroying the very fabric of the world's environment in the process. Prophetic preaching about a topic like this can be very difficult in specific situations. What, for instance, is a priest going to say about the issue of environmental destruction in an area that depends on the export of woodchip for its livelihood? Should he go out on a confrontational limb and run the risk of alienating many of the congregation?

As I learned some years back, there is no easy answer to this kind of question. I once worked in a parish just outside Boston, Massachusetts, where nearly all the parishoners were involved at a managerial or research level in the electronic weapons industry. It was at the time the United States bishops were evolving their now famous pastoral letter *The Challenge of Peace* by widespread consultation. [15] The parish priest, who was a most intelligent and perceptive man, felt that simply confronting these people with the moral reprehensibility of almost all modern weapons would merely create useless tension in the parish. The family lives of these people were caught up with the weapons corporations. In the US, the

---

15. For a description of the process used see Castelli, Jim, *The Bishops and the Bomb: Waging Peace in a Nuclear Age*, Image Books, New York, 1983.

electronics industry is very integrated and interconnected; to walk out of making weapons is to walk out of the industry altogether; the person who does so will never be employed again in the one area in which they have been trained. So what do people do whose whole family lives are tied to a specific area and industry through social networks and financial commitments? Usually, in preaching, it is not the priest's job that is on the line. Nevertheless what can a local priest say in such a situation? Many would opt for compromise or silence — but this can not be the answer. The priest on the spot has to find a position between compromise and useless ranting. It involves sophistication, patience and a sense of nuance. It involves the long-term process of the formation of adult conscience. This will involve the breaking down of the schizophrenic split which characterises the lives of many immature Catholics, between Sunday religion and everyday life. They need to be helped to see that Christian moral and religious commitments must pervade the totality of life, including their working life. There is a need for ongoing education in the church's teaching on morality, including the pastoral application of moral principles. Finally, the parish community as a group is going to have to confront the difficult practical task of helping people to evolve other, more moral options for themselves. The discussion of the prophetic aspect of preaching brings us to the vital need for on-going adult formation in faith. [16]

The greatest strength that the preacher has is that God's Spirit entrusts to him the task of uttering the word of God. Karl Rahner, in his brilliant essay 'Priest and Poet', argues that the word of God only becomes freed and actualised when it is interpreted by the preacher. [17] The preacher's task, Rahner says, is to free the word of God. In order to do this the preacher himself must have been converted by the very word he utters. When Rahner speaks of 'freeing' and 'actualising' the word, he means that the preacher must bring to consciousness the rich, latent truth in the biblical word that will touch contemporary men and women. To do this the preacher must use what Rahner calls 'primordial words'. Such words, he says

are like sea-shells, in which can be heard the sound of the ocean of infinity, no matter how small they are in themselves. They bring light to *us*, not we

---

16. See pp. 100–106.
17. Rahner, Karl, *Theological Investigations: The Theology of the Spiritual Life*, Darton, Longman and Todd, London, 1967, Vol. III, pp. 294–317.

to them. They have power over us because they are gifts of God, not creations of men, even though perhaps they come to us through men. Some words are clear because they are shallow and without mystery . . . Other words are perhaps obscure because they evoke the blinding mystery of things. They pour out of the heart and sound forth in hymns . . . Such words, which spring up out of the heart, which hold us in their power, which enchant us, the glorifying heaven-sent words, I should like to call primordial words. [18]

For Rahner the task of the preacher is even more important than that of the artist, for the preacher is asked to speak words which open deep and transcendant realities for the participating congregation.

To achieve this symbolic, sacramental task, the preacher must be immersed in two worlds: that of the Christ of history and faith and that of the present day. In the best sense the preacher must be both a person of faith and a man of the world. He must draw from his own faith, conviction and spiritual experience to enlighten everyday reality. There are very few good sermons, I suspect, because there are very few preachers who have lived life fully and reflected on their lives in the perspective of God's word. For an effective sermon there has to be a sympathy with contemporary life. Sermons so often are spiritually dead, because the preacher has never come to life through living the tension of faith and experience.

From this it is clear that only specifically gifted people within the church community should be allowed to preach. So far I have referred to preachers in the masculine, because it is generally presumed that it is only priests who are canonically permitted to preach in the Catholic church. However, this has not always been true. Father Edward Schillebeeckx points out that it was only in the Middle Ages that lay people were banned from preaching. [19] For example, there is 1000 years of tradition of preaching by monks (who were still laymen right up to the Middle Ages). It was lay Irish monks who converted important regions of Europe through their preaching in the 7th and 8th centuries. Saint Francis of Assisi himself, who founded the Franciscans as itinerant lay preachers, was only ordained near the end of his life as a deacon. Even in our own time, canonist, Father James Provost of the Catholic University of America, points out that

---

18. Rahner, op. cit., pp. 295–296.
19. Schillebeeckx, Edward, 'The Right of Every Christian to Speak in the Light of Evangelical Experience "In the Midst of Brothers and Sisters"' in Foley Nadine (ed), *Preaching and the Non-Ordained: An Interdisciplinary Study*, Liturgical Press, Collegeville, Minn., 1983, pp. 11–37.

there are 'cautious openings to lay preaching present in the revised Code [of Canon Law]'. [20] It is becoming increasingly clear that the church will have to move to a situation where the person (either man or woman) who is gifted with biblical, spiritual and human insight and the gift of communicating those insights, will be anointed (or, as I have already suggested 'ordained' [21]) to preach the word to the congregation. The close nexus between the celebration of Mass and the sermon will probably be broken down and popular preaching will be again used as widely as it was in the Middle Ages.

## The parish of the future

This discussion of the renewal of worship and preaching brings us back to the structure within which most Australian Catholics experience the church — the parish. What is the future of the contemporary parish? And what is happening to the changing relationships of the people who primarily make up the parish — the laity and the clergy? Two fundamental movements are occurring that are going to determine the future of parishes in Australia for a long time to come: negatively, there is the shortage of priests and their increasing average age, and positively there is the new emphasis on the importance of lay ministry in the contemporary life of the church.

This shift in the life of the church throws up a basic question:

What sort of parishes do we want in ten years time? And what sort of decisions can be made in the coming years to ensure that viable parishes — parishes with adequate members, structure, staff, programmes and finances -are maintained. [22]

Throughout 1990 the Catholic Research Office for Pastoral Planning in Melbourne set up a consultative process in the deaneries of the archdiocese on the future of parishes. Because it is clear that this consultation is typical, at least, of urban dioceses in Australia, I think that it is worth examining closely. It is true that only a very small group of people out of the Catholic population of the archdiocese participated in the consultation, but this does not invalidate the

---

20. Provost, James, 'Lay Preaching and Canon Law in a Time of Transition' in Foley, Nadine (ed), op. cit., p. 152.
21. See pp. 43–44.
22. *Cropp News*, August 1990, p. 5.

conclusions. In the process of change it is élites who enunciate new ideas and make the first moves. The people who participated in the Melbourne exercise are the ones who will influence others, who will themselves work in the ministry of the church and give currency to new ideas by taking them back into the mainstream.

In the light of the increasing shortage of clergy in Melbourne, the Research Office proposed four possible models for the future of parishes:

(1) The local parish continues in existence as it is now, but with a non-ordained pastoral administrator. This option has, for example, already been taken up by a considerable number of rural parishes in the United States and some parishes in Tasmania, Victoria, Queensland and Western Australia. For instance, in the Western District parish of Harrow, in the Diocese of Ballarat, a Sister of Mercy, Kathleen Moran, acts as pastoral administrator. She describes her role this way: 'I do not see myself as a priest substitute but as one of the community, in a leadership position, helping others to live out and recognise their pastoral calling'.

(2) The local parish continues with a part-time parish priest (who has specialist duties elsewhere) who is responsible for the celebration of the eucharist, with or without a resident pastoral associate. The Melbourne parish of Prahran has taken up this option; the parish priest doubles as chaplain to the deaf and Presentation Sister, Michelle Kennan, co-ordinates the parish. [23]

(3) The local parish continues in a modified form through amalgamation or 'twinning' with another parish and a priest is shared between the two parishes. There are a number of examples of this in Australian rural parishes.

(4) The parish is closed down and the priest withdrawn. In many cases the church property is sold.

In all consultations of laity in Melbourne, the option of closing the parish down was decisively rejected. The consensus was that if there is a viable local community, the parish should not be closed. The problem was the definition of the word 'viable'. One of the Melbourne deaneries suggested that 'Criteria for viability be established, and (that) all parish boundaries should be reviewed regularly'. The closure of parishes in decayed areas of the inner cities in the United States (such as in Detroit and Chicago) has led to widespread protest and dislocation in these archdioceses. In Detroit the determination of Cardinal Edmund Szoka to close a large number of inner city parishes, because of a lack of financial viability, led to accusations that the archdiocese was 'abandoning the poor in the inner city' and

---

23. See *The Advocate*, 14 June 1990.

to a protracted public confrontation between the archbishop and parishioners. Many of these parishes were in former ethnic ghettoes with large churches and extensive plant. In Detroit they were eventually closed, but it was followed by Szoka being moved from Detroit to a job in the Vatican and to the need for a long-term healing process in what was formerly an excellent and ministerially creative archdiocese.

In the Melbourne consultation, the option of closing down parishes was decisively rejected by the deanery meetings of laity and clergy. This is clear in the reports from deaneries:

There was a very strong feeling that wherever possible local communities should be preserved. This meant that in general that the option of closing parishes was firmly rejected.

The reason for this is clear. For many Catholics in urban areas, the parish has a primary mediating function; it is a place where people can meet others they know, or even if they do not know them, they are familiar strangers with whom they have worshipped for a considerable time and in whose company they are comfortable. It is an example of commonality actually in operation.

Several other issues emerged from the Melbourne consultation. For most Catholics, the leadership role of the priest is pivotal. This seems to arise from a deep, instinctive Catholic recognition that priestly leadership, especially in the celebration of the eucharist, is central in the church's tradition. The evidence from the consultation is that priests, whether married or celibate, whether men or women, will continue to exercise a key sacramental and leadership role in the church. The need for priestly leadership was put forcefully:

The importance of the ordained ministry was recognised in all responses as was the essential link between priesthood and eucharist. The priest's major role is sacramental and spiritual. The priest is minister, leader, nurturer and innovator. There is a need for research to determine the most appropriate style of priesthood for the future (Eastern Region).

The continuing need for a symbol of leadership is inescapable in any institution like the church. Absolute democracy and total equality is simply not possible even in human affairs, let alone in church affairs. The key element in the church's understanding of itself is diversity of gifts drawn together by the Spirit of God.

Another striking element in the reports of the Melbourne

consultation was the openness of the laity to married clergy, the return of resigned priests and the ordination of women. For instance, the Southern Region said bluntly: 'The voices of many people are saying: consider the ordination of married men and of women and consider recalling resigned priests to active ministry'. The majority of Catholics do not perceive any obstacle to the broadening of the basis for recruitment to the priesthood. And while I am not saying that public opinion determines the truth or falsehood of a position, there is such a thing as 'consulting the laity in matters of doctrine'. The phrase is the title of a famous essay by John Henry Newman who is clear as to why the laity should be consulted: 'The body of the faithful is one of the witnesses to the fact of the tradition of revealed doctrine, and because their consensus through Christendom is the voice of the infallible church'. [24] It is a pity that some bishops do not take Newman's advice.

Another other key issue that emerges from the consultation is the role of laity in ministry. I use the word 'ministry' here purposely. There has been an attempt on the part of some reactionaries in the church to limit this word to the work of priests. This debate over ministry has been part of the convoluted and continuing criticism by the Congregation of the Doctrine of the Faith of Dutch Dominican theologian, Father Edward Schillebeeckx. [25] While the criticism may be making life uncomfortable for Schillebeeckx, it is clear that Rome is losing this battle on the wider front by default. Lay people are not only talking about 'ministry', they are actually engaged in it and many priests and bishops are supporting them in the process. Without lay people, the ministry of the church would collapse in most parts of the world. There is no doubt in the minds of those who participated in the Melbourne Consultation that laity are called to ministry by virtue of their baptism and confirmation.

The present time was seen as one of great opportunity and challenge for lay people in the church. The promotion of lay ministries was seen to be

---

24. For the argument that surrounded the composition of this essay see Ker, Ian, *John Henry Newman: A Biography*, Oxford University Press, Oxford, 1988, pp. 480–483,
25. See Willems, Ad, 'The Endless Case of Edward Schillebeeckx' in Küng, Hans, & Swidler, Leonard, op. cit., pp. 218–222. It was Schillebeeckx's book *Ministry: Leadership in the Community of Jesus Christ* (Crossroad, New York, 1981) that was targeted by the Roman Congregation of the Doctrine of the Faith. He has subsequently published *The Church with a Human Face: A New and Expanded Theology of Ministry*, Crossroad, New York, 1990.

important, not only because of declining numbers of priests but as the right and duty of lay people by virtue of their baptism. Even if numbers of priests were not declining, greater lay involvement would still be necessary to fully develop the potential of the parish to meet the needs of its members (Northern Region).

One of the central issues that results from lay involvement in ministry is the need for training. This was made clear in the Southern Region of the Melbourne Consultation: 'Lay theological study and adult faith education should continue to be strongly encouraged so as to ensure that sufficient lay people are prepared for ministry and leadership in the future'. The church's ministry could face the danger of being taken over by untrained and inept amateurs, who will make up for a lack of sophistication with an overdose of dogmatic self-righteousness and uncompromising certainty.

The Australian Catholic church has an extraordinary number of laity involved in some form of formation in faith or in programmes of adult growth. Sister Roberta Hakendorf claims that 'a lay revolution looms in the Catholic church'. [26] This has been one of the great unspoken success stories of the local church. I emphasise the word 'formation' here, for what is happening is something much more than what used to be called 'adult education'. The emphasis is not only on the development of knowledge about religion and theology but also on a living experience of faith, both in its communal and personal dimensions. The aim is to move adults from the stance of passive spectators in the church toward involvement and ministerial action appropriate to their overall commitments.

The sheer number of people involved indicates how deeply adult formation has permeated the Catholic church. The Sydney Archdiocesan Lenten Programme has now been going for twelve years and in the Lent of 1991 more than 50 000 people worked through the programme. The *Renew* programme (despite a sustained sniping campaign from reactionaries) has already been completed in the Melbourne Archdiocese, where 18 000 people worked through a three year long programme, as well as in the dioceses of Cairns, Armidale, Maitland, Wagga, Port Pirie, Sale, Ballarat and Sandhurst (Bendigo). In 1991, the archdiocese of Hobart and the dioceses of Rockhampton, Townsville, Parramatta and Bathurst were still working through the

---

26. Hakendorf, Roberta, 'A lay revolution looms in the Catholic church', *National Outlook*, November 1988, pp. 15–18.

programme. Australia-wide, close to 100 000 people have been through the *Renew* process.

The history of adult formation in Melbourne Catholicism goes back to the 1930s to the Campion Society, and beyond that to the Legion of Mary and the Catholic Young Men's Society (CYMS). Emerging also in the 1930s and the 1940s were the Young Christian Workers (YCW) and the National Catholic Girls Movement (NCGM). The 'Movement' (the Catholic Social Studies Movement), the National Catholic Rural Movement and the closely connected Institute of Social Order (a Jesuit-run organisation) gave many lay people a sense of participation in the life of the church, although the Movement eventually became subsumed to political priorities. Adult education in Sydney goes back to the 1940s. Roy Boylan, Mary Gilchrist and the Paulians were the pioneers. In Adelaide the Newman Institute of Christian Studies (begun by Bill Byrne and Ted Farrell) lasted from the 1940s to the 1960s. In Perth there was the Catholic Social Apostolate. The theology of the lay apostolate was given a great boost in 1957 when the French Dominican, Yves Congar, published the English translation of his seminal theological study on the laity. It was to have a profound and continuing influence. [27]

While there is real continuity with these early initiatives, the present adult formation movement really began in the years during and immediately after Vatican II. Part of it was involved with the spread of the Christian Family Movement and organisations like the Paulians and the Christian Life Federation. In education during the late 1960s and the early 1970s there were the catechist courses that helped the lay people engaged in religious instruction of Catholic pupils in state schools. But in some ways it was the International Eucharistic Congress in Melbourne in 1973 which really got things going. It was preceded by a renewal programme which focused on eucharistic theology and justice questions and was designed for an ecumenical setting. By the mid 1970s, the adult formation movement was nurtured and guided by far-sighted people like Fathers Noel Molloy and Tony Doherty and Sister Christine Burke in Sydney, Peter Gagen in Brisbane, Kevin Treston on a national level, Father Dennis Edwards and David Shinnick in Adelaide and Father Peter Nicholson in Melbourne. One of the guiding pioneers was also Monsignor John F. Kelly who, even before Vatican II, had prepared the *Australian*

---

27. Congar, Yves M. J., *Lay People in the Church: A Study for a Theology of Laity*, Geoffrey Chapman, London, 1957 (Engl trans).

*Catechism* (which replaced the famous old 'Penny Catechism') and who in 1973 became first Director of the National Pastoral Institute in Melbourne.

In the 1970s the philosophy of the adult education movement changed under the influence of the American thinker Malcolm Knowles. The focus shifted from an instructional methodology modelled on schooling, to a focus on achieving adult attitudinal change. A key document in Australia in the evolution of Catholic adult formation was the 1983 publication *Towards Adult Faith*,[28] prepared by the Catholic Adult Education Association. The booklet describes the development of the Association:

In August 1979, a group of men and women from throughout Australia gathered at a conference centre between Wagga and Junee. All of us were working full-time in the field of adult education in faith, mostly within diocesan offices. Many of us had been in contact before . . . we elected to begin to organise ourselves into a formal structure to be known as the Catholic Adult Education Association (p 3).

The booklet points out that adults were unprepared for the changes that followed Vatican II, especially when apparently 'unchangeable' and 'age old' teachings, beliefs and practices were discovered to date merely from the 16th century — or even later. It argued that the faith education of adults was the 'critically important pastoral priority' that required 'the same courage, resourcefulness and intelligence which was marshalled around the education of the young a century ago' (p 62). It then outlined the major issues that the association said should underpin and be confronted in any process of Catholic adult formation.[29] I will list these issues in detail because they set out an agenda for adult formation in the church of the coming decade:

(1) How do you motivate people to see the value of continuing faith education beyond school days, and of the need to explore as adults the relevance of Jesus and his teaching?

(2) How do you help people discern the call of Jesus in the losses and breakdowns that characterise modern life e.g. marriage breakdown, unemployment, the loss of loved ones?

---

28. Catholic Adult Education Association: *Towards Adult Faith: Adult Education in Faith in the Catholic Church in Australia Today*, 1983.
29. *Towards Adult Faith*, pp. 64–69.

(3) How do you deal with the loneliness and alienation of modern life? How much help is it to belong to a group of believers?

(4) What are the two-thirds of non-practising Catholics telling the church community and how can these people be welcomed back into new experiences in a changing church?

(5) What do Australians from a non-Anglo background have to teach the Australian church? What of Aboriginal Australians? Does Catholicism make any sense to them?

(6) Given that most Catholics now belong to the middle or upper middle class, how can the church be a church of the poor?

(7) What does 'action for justice' mean for Australian Catholics? How do you build a more just society? How do you challenge the apathy of the majority of Catholics?

(8) How does the church develop new styles of ministry and leadership, given the shortage of clergy? What will be the pattern of leadership in the church in the year 2000?

(9) How does the church integrate women into all pastoral areas, including opportunities for leadership?

(10) How do you get pastors and parish councils to make adult formation in faith a key priority? What resources will they need? What resources will dioceses give to adult faith education?

(11) How can the ability of the Catholics who work in the media be used effectively by the church? What vision of the church is conveyed by the Catholic press?

In 1990 these key issues were updated and the need for a theology of the environment and a response to fundamentalism were added to the list.

What is 'Catholic adult formation'? People who have grown up with a Catholic schooling often think that lay formation is 'getting more information', or learning 'new approaches' to faith in a cognitive sense. Many of the Australian bishops apparently also tend to think this way. The cognitive is certainly an element in the process, but, as the director of the Sydney Catholic Adult Education Centre, Father Tony Doherty, argues, adult formation is not merely the absorption of information through lectures and classes, but is a process of trying to get people to move from their passive, assumed and often unconscious certainties and fixed attitudes to involving themselves and participating in something new. Once they begin to act they become more vulnerable; the educational moment is actually in that movement toward action. In other words, if you can change people's behaviour from passive to active, things begin to happen. The Sydney archdiocesan programme *Parish 2000* (now also used in some parishes in the New South Wales dioceses of Broken Bay, Wollongong and Wilcannia-Forbes) is an example of what he is talking of. People are involved by asking them to help organise or do things for the programme; once they begin acting, they become open to change.

Doherty also expresses concern about the fundamentalism that characterises so many Catholics. This is not just manifested in biblical fundamentalism, but in an excessively simplistic understanding of religion. It can be dealt with not only by further education in the complexity of biblical and religious studies but, more importantly, by getting people to change their behaviour by challenging them to act as adults, and in the process to confront their fundamentalisms.

The most interesting thing that strikes you when you examine the programmes around Australia is that all of the major metropolitan archdioceses have developed different styles of approach to adult formation. All are valid and all have worked well within their contexts. The Sydney Centre, for example, also runs Parish Ministry Publications among whose projects is the very successful *Four Gospels* by Father Michael Fallon. Brisbane has a Faith Education Institute, and Melbourne's Catholic Pastoral Formation Centre offers a range of courses in basic theology, biblical studies, sacramental programmes, pastoral planning, developing small Christian communities, pastoral ministry, communication skills for ministry, pastoral liturgy and group and helping skills. Adelaide has a Catholic Adult Education Service which offers extensive courses similar to those offered in Melbourne. It also runs a two year leadership programme which aims to give participants an introduction to theology, biblical studies, sociology and formation in faith, to prepare them as leaders and to help them develop a sense of an effective Christian presence in their workplaces.

## Lay leaders in the church

One of the key issues facing the church today is the development and formation of lay leadership. As the priest shortage becomes more acute, the question emerges as to how adult formation programmes will train laity to undertake a leadership style that is open and participative. The thing that has to be avoided is to prevent the laity becoming a cardre of 'junior clerics', often narrower than the priestly clerics they replace. That is why I think the Australian bishops, unlike the US episcopal conference, have been wise not to introduce the permanent deaconate into Australia. I also think that those dioceses that introduced male acolytes (adult altar servers) have probably made a mistake. Many of the men who join the ranks of the acolytes often easily and happily become very clericalised.

The archdiocese of Adelaide has tackled the issue of lay leadership formation strongly influenced by the thought of Belgian Cardinal Joseph Cardijn; the YCW and the YCS were always strong in Adelaide. [30] For many years Adelaide had the Christian Life Movement which was aimed specifically at lay formation along the lines of the spirituality of the YCW — 'see, judge, act'. The Movement is now integrated into the Christian Community Development Service which promotes leadership formation. Also linked is the Diocesan Movement for Small Christian Community. These are not inward-looking groups concerned with self-engrossed navel gazing and 'warm' feeling with others, but are geared to look outward towards the needs of the world. Their aim is to build a Christian community *for* the world — a lot different from simply building a Christian community *in* the world.

A special problem in many parts of Australia is distance. How do you get people together, especially in a rural diocese, for personal formation? Even in the cities, people need to be with their families and often cannot find time to be out. Brisbane's Faith Education Services is trying to confront these problems. They have developed a Christian Leadership Programme which involves a strong element of the spirituality of leadership. The programme is set out in booklets that integrate both text and response and structured so as to develop a dialogue with the learner. Participants do not have to leave home to do it and they come together in local cluster groups only periodically. Some of the dioceses with vast areas have replaced meeting with teleconferencing. The Christian Leadership Programme is operating in all Queensland dioceses, as well as Armidale, Darwin and Port Pirie, and it involves people working in schools, health care and parishes.

The Melbourne archdiocese committed itself to the *Renew* programme in 1986. The programme originated in the US archdiocese of Newark, New Jersey, in 1978. Despite reactionary opposition,

---

30. The YCW (Young Christian Workers) and YCS (Young Christian Students) actually began their activities in Australia in 1941 under the patronage of the former Archbishop of Adelaide, Matthew Beovich.

the Australian version of the programme has been very successful; close to 100 000 people have passed through the three year process. It is designed as a grass-roots programme, based on parishes, that helps people understand what is going on around them in contemporary Australia and assists them in integrating their Christian life into that. It is a systematic process of spiritual renewal that helps growth in prayer, understanding the scriptures and the formation of living faith communities and helps people appreciate justice as an integral part of Christian living. [31] It has certainly helped many people who had really not understood nor integrated what had happened in the church in the years after Vatican II.

One of the most radical documents resulting from Vatican II was issued by the post-conciliar Commission on the Liturgy on 6 January 1972 and entitled the *Rite of Christian Initiation of Adults* (RCIA). [32] The RCIA is not just a restoration of the old catechumenate, the initiation process of the ancient church for converts. It is a modern adaptation of that ancient rite that was first revived in Paris in the 1940s and whose initial modern development was in the francophone world.

The RCIA set up a new formation process for those who want to become Catholics, which involves liturgical, communal and instructional elements — adults who make the decision to become Catholics enter into a process of formation in faith. The RCIA emphasises that this process is not just a matter of what catechumens (the initiates) believe, or a question of their personal choice, or of being a 'convert'. Rather it emphasises the development of new attitudes, involving the whole local Catholic community in the formation of catechumens:

The formation of the catechumens is not to be merely intellectual but a matter of living. Their deeper knowledge of God and Christ must be accompanied by an ongoing effort at the renewal of their lives . . . the Christian community also plays an important and irreplaceable part; it

---

31. See the booklet *Renew Australia: Why, What, How? A detailed overview of the RENEW process*, 1986, pp. 7–15.
32. Archbishop Annibale Bugnini, the key person in the entire liturgical reform of Vatican II, tells the story of the *Christian Initiation* document in his book *The Reform of the Liturgy 1948–1975* (Liturgical Press, Collegeville, Minn., 1990, pp. 584–597). For a summary of the history of the adult catechumenate and its modern revival see *Becoming a Catholic Christian: A Symposium on Christian Initiation*, Sadlier, New York, 1978, pp. 8–34.

actively shares in the celebrations, concerns itself with the candidates and joins them in an effort at conversion and renewal. [33]

The catechumenate began in practice in Australia in 1979 (although a small group of people in Australia had looked closely at it for a number of years. I had introduced students for priesthood to it in Saint Paul's National Seminary in 1974; a number of other seminaries also dealt with it with around that time). The RCIA has been successful and in 1991 more than 1000 people entered the church at Easter time, having followed the months-long RCIA *Journey to Easter* programme. Both Melbourne and Perth archdioceses have catechumenate offices. Linked to the catechumenate is the *Re-Membering Church* programme, which helps lapsed Catholics return to the practice of Catholicism. Recently, the Brisbane *Catholic Leader* profiled the typical Australian catechumen:

More [candidates] have approached the parish or diocesan office about the programme spontaneously, are unbaptised, and there are many more Asian candidates in some dioceses. He or she — and now it is more likely to be *he* — averages about 30 years of age, although there are candidates from early teens to the seventies, and children among whole families who are coming into the Church. They will come from occupations across the board, but there seem now to be more with tertiary education. [34]

The RCIA is going to become a permanent part of the liturgy of the church and I would predict that there will be a continuing and increasing flow of adult converts to Catholicism. Most of these will be unbaptised and will have grown up in a non-Christian family. The Catholic church will increasingly appeal to well-educated people, so it is important that the RCIA not neglect the need of these people for a coherent apologetic for Catholic Christianity.

## Formal or informal ministry

There is a common problem participants face at the end of all of these programmes. It is: Where do people go with the formation they have received? What do they do with what has happened to them? It is a very difficult process to capture all this energy and transfer it to a parish, a school or other church structures. Often these structures are unyielding and unresponsive and still working

---

33. Bugnini, op. cit., p. 592.
34. *The Catholic Leader*, 31 March 1991, p. 3.

out of a concept of the church that sees the laity as passive recipients rather than as active ministers. After the *Renew* programme was completed in Melbourne, an effort was made to channel the energy into parish planning and leadership and several parishes have begun to re-structure themselves in terms of small communities. But these are very much the exceptions.

A similar problem arises with the enthusiastic movements like Cursillo, Marriage Encounter, Antioch and Teams of Our Lady. People move through these programmes and emerge with an articulated, adult faith, but how do they move on to ministry, social action and the transformation of society? How is this achieved in a church whose structures are only changing very slowly? How do you maintain a living worship and some form of spiritual support for those who are 'all dressed up and ready to go'? This is the key question that now confronts the leading people in the adult formation business.

There is no doubt that it is an urgent and important question, but I think it is also a bit misleading. It is unreal to expect that everyone who has gone through some form of renewal will want to or even be able to assume a ministry or a leadership role. There are many committed Catholics who feel that their commitment is lived out in the context of their family, their participation in the life of the church and their work in society. It is also imperative that people be given the chance to move in and out of ministry according to both their external circumstances and their personal development and needs. Family and other commitments often mean that they cannot do both a fulltime job and work in ministry.

Also, as the Adelaide renewal emphasis makes clear, the primary focus of ministry should be outward, towards the world. Most lay ministerial energy ought to be directed toward permeating the society with the spirit of Christ, and attempting to deal with the complexity of changing unjust and exploitative societal structures. For many lay people this will be achieved in and through their everyday lives and not in some special ministerial role. The church exists for the sake of the world, not for its own sake. As Catholics gain more power and influence in the various levels of Australian society, it is imperative that they are formed to begin to use that power on behalf of those in our society who are exploited and have no voice. To be a Catholic these days demands a commitment to justice. The leaders of the church cannot allow the power brokers who belong to the church to continue to indulge in the traditional Catholic schizophrenic split between pious Sunday religion and morally neutral or exploitative or even morally reprehensible weekday behaviour. These people ought to use their influence to build a more just society.

Only a small percentage of lay people will ever work fulltime and directly in the ministry of the church. One of the best examples now of fulltime lay ministry is the pastoral associate programme. Already these are more than 350 pastoral associates at work in 21 dioceses in Australia. This ministry evolved from the parish catechetical classes and motor missions of the religious orders. The first systematic approach to catechetics in isolated rural areas was devised by the Missionary Sisters of Service in Tasmania. Many other religious sisters began to take up this role, often working where there was no priest. The idea of a pastoral associate evolved out of this and the first of them were appointed in the mid-1970s. As a result of their origin, most pastoral associates are still members of religious orders. The Melbourne archdiocese has the most highly developed system of pastoral associates and in 1990 the Catholic Research Office for Pastoral Planning examined in detail the experience of twelve accredited lay pastoral associates. In Melbourne at present:

there are about 160 pastoral associates working in parishes in . . . [the] Archdiocese and a further twenty or so are engaged in hospital work. Of the 160 parish workers, just over one hundred are religious sisters, four are religious brothers and about fifty are lay people. [35]

They work in over half the parishes of the archdiocese. Their backgrounds are varied as is their pastoral work. It includes 'liturgy preparation, preparing couples for marriage, visiting the sick and elderly at home and in hospital, planning and presenting adult education programs, being involved in diocesan, ecumenical or local affairs, supporting the RE teachers and the RE co-ordinator in the parish school'. [36] The research study found that pastoral associates saw themselves as leaders in the parish and as responsible for empowering other lay people. The women among them commented that they hoped to bring a feminine perspective to leadership: 'There is something about a woman in terms of vision and compassion and openness and heart-level operating that is absolutely essential to

---

35. *Cropp News*, November 1990, p. 1. Most of the issue reports on the Melbourne research project which was carried out by Robert Dixon.
36. *Cropp News.* art. cit., p. 3.

Christian leadership'. They generally reported high level job satisfaction and most did it because they had a sense of call or vocation, or of a job needing to be done. 'In most cases this was expressed in terms of forming or belonging to a church community, or to taking up the challenge of lay or female leadership'. Most salaries were worked out between the associate and the parish priest and 'ranged from $15 500 to $26 000'.

There is no doubt that the pastoral associate model will spread and develop right across Australia and it could prove to be a very flexible model of ministry. Pastoral associates will not carry the heavy tradition attached to the priesthood and, as local people, they will probably be better attuned to the needs of the area in which they are working. It is from the ranks of the lay pastoral associates that a married clergy will probably emerge.

There is a danger, however, that the pastoral associate model will become 'clericalised'. The best guard against this is that parishes always maintain a mix of professionals and volunteers. A natural human tendency is to hand over everything to the professionals; people often argue 'They know what they are doing'. But if there is a strong emphasis on volunteers, especially trained ones, then the natural clericalising tendency of professionals will be resisted.

These various *caveats* should not obscure the success that has already been achieved in developing lay ministry. The Australian Catholic church has a corps of laity ready for ministry; many of whom are already partially involved in ministry because, almost despite itself, the church is actually engaged in the process of developing, willy-nilly, a plurality of ministries. It is a good example of Australian pragmatism: without any over-all planning, the church actually finds itself ready to replace the shortage of clergy with willing laity. I am optimistic that the church will not miss this opportunity to grow and develop.

## *Catholic schools*

The most radical and prophetic action that the Australian Catholic bishops have ever taken was the decision in the 1860s and 1870s to set up a school system. True to form, however, they had to be pushed to make a cohesive decision about the whole thing — although, in their defense, it has to be said that the pace of educational change in each of the different colonies was different. From about the middle of the 19th century, the Catholic bishops were confronted by a rampant and agressive secularism that was determined to wrest control of education from the churches. The bishops' responses were determined to a large extent by the different circumstances in each of the different colonies. [37] The most prophetic Catholic leaders against this secularist push came from South Australia. The key bishops were Bonaventure Geoghegan (d.1864) and Laurence Sheil (d.1872) of Adelaide. Advising Sheil was one of the greatest and most far-sighted priests of 19th century Australia, Father Julian Tenison Woods. It was Geoghegan, however, who first advocated 'a complete break with the state and the establishment of an entirely independent Catholic system'. [38]

By 1869 the bishops were ready to make a stand. They were alone, as all the other churches surrendered their rights in education to the state, although significantly Anglicans and Protestants held on to their more exclusive private schools. In order to make their stand a reality, the bishops had to set up a primary education system from scratch. They achieved this principally through the quick importation of religious teaching orders of nuns and brothers from Europe and especially from Ireland. As I said in *Mixed Blessings*:

The bishops simply refused to compromise on fundamentals. For them the basic principle was that Catholic Christianity must permeate all education. Certainly their motives were not totally religious or disinterested. Some of them were spoiling for a fight! But their action does have a prophetic ring. They contended that an explicitly Catholic ethos must pervade the whole of education. [39]

Some of the inspiration for this stance came from the papacy and from the rejection of modern liberal society by Pope Pius IX. Certainly, from the education crisis onwards, the theology and ecclesiology underpinning the bishops' attitudes placed the Australian church over and against liberal 19th century society and Catholics began to picture the church as the one bastion against the erosion in society of Christian doctrine and values. In some ways, it was the beginning of the Catholic 'ghetto' and of a sub-cultural attitude.

Despite the bishops' ideology, the decision to set up a school system

---

37. For a treatment of the whole affair see Fogarty, Ronald, *Catholic Education in Australia 1806–1950*, Melbourne University Press, Vol I, pp. 170ff.
38. ibid., p. 172.
39. *Mixed Blessings*, p. 219. See pp. 216–220 for treatment of the education question.

was truly prophetic. They knew that 'the time had come' to make a stand and they were prepared to settle their not inconsiderable internecine and personal differences to show a united front.

## The change in Catholic education

In my view, Catholic education faces a similar prophetic watershed today. It is not about the survival of the Catholic system; that is assured. It is about the *purpose* and *direction* of that system. Extraordinary changes have occurred in all Catholic schools over the last thirty years. We have come from a situation of absolutely no government funds for Catholic schools, to a very high level of funding. The symbolic turning point was the Goulburn 'strike' of 1962, when Catholic parents threatened in the course of one week to close all the Catholic schools in a city with a very large Catholic minority and to demand admission for the children to the state schools. It was a potent gesture that thoroughly frightened the New South Wales state government and Catholics should not forget its power as a symbol.

The strike was thirty years ago. Personally, I remember it well; I was a philosophy student in the seminary in Canberra at the time; and Goulburn was literally just up the road. Significantly, it also happened during the episcopate of Archbishop Eris O'Brien of Canberra and Goulburn, one of Australia's best historians and an excellent bishop, with wide contacts at all levels of Australian society and especially with the federal government. (There is an old story in Canberra that in the 'early days' many senior government people and politicians, including Prime Minister Menzies, often used to visit Archbishop O'Brien, because it was the only place where you could get a decent meal in Canberra! It may be apocryphal, but it indicates something of the closeness of Dr O'Brien to the government of the day.) As Bede Nairn, former editor of the *Australian Dictionary of Biography*, comments:

[Archbishop O'Brien] was very much aware of the great post-war strain on the Catholic school system and in 1962 he gave his support to the temporary closure of schools in Goulburn. This incident supplemented the wise affability with which he had negotiated with the Commonwealth Government for financial aid, and in 1965, a degree of capital assistance was given in the Australian Capital Territory. This decision began a national trend. [40]

---

40. Nairn, Bede, 'Obituary — Dr Elis O'Brien, P.P., M.A., Ph.D., *Journal of the Australian Catholic Historical Society*, 12(1990), p. 24.

The Catholic system is now deeply dependent on that government aid.

Another major change in Catholic education has been the shift from a predominantly religious staff in schools to a predominantly lay staff. In thirty years the Catholic community has seen its schools move from ninety per cent staffing by religious sisters, brothers and priests to more than ninety per cent staffing by lay people. Many religious orders have already surrendered control of their schools to lay teachers or are planning to do so. As a result of this shift away from the control of the religious orders, most of the schools have become much more 'systemic' in that they have considerable control imputs from a centralised diocesan bureaucracy. No longer are schools run by this or that religious order, but by diocesan educational authorities. Catholic education has thus become much more centralised.

There has been very little philosophical reflection on this profound shift in emphasis in Catholic education. The method of change has been both very Australian and very Catholic: it has happened almost totally ad hoc. There has been little or no set agenda and much of what has happened is as a response to ever-increasing government funding and major changes in the structure of religious orders. But this ad hoc response cannot continue. The Catholic community is going to have to decide what it is going to do with this extraordinary ministerial resource. As the church moves toward the third millenium it has to decide on a new direction in education, especially given the involvement of so many lay teachers. I would argue that the key to the future lies in a reassessment of what the church understands by ministry and how the ministry of the school fits into the overall work of the church.

## Offering an alternative

Throughout this century Australian Catholics have argued for their schools on the grounds that they offered an alternative to the secular philosophy of the state system. This word 'alternative' is the one of the keys to the Catholic philosophy of education. But the Australian church has still not articulated what it means by 'alternative'. One of the meanings of the word offered by the *Macquarie Dictionary* is fascinating in terms of Catholic education: 'Offering standards and criteria of behaviour of a minority group within and opposed to an established western society'. In other words 'alternative' means that Catholics, albeit a rather large minority group in Australia, are supposed to be offering something different to the explicit secularism of state schools. It is interesting in this context that, while the state system has never repudiated its secularist basis, it *also* lacks a sound philosophical base. The one catchcry the state schools have seems to be 'equality of opportunity', although even this philosophy of education has also been insufficiently articulated and developed by its proponents. Secularism today has largely degenerated into crass materialism, although many state schools still struggle to give a moral basis to their educational philosophy.

The Catholic church is better placed to develop its own coherent philosophy of education. It certainly cannot be just a copy of the state system with religious education added. 'Alternative', in the Australian context, means that the Catholic system must attempt to offer a coherent and explicitly different Christian set of beliefs and values which are rooted in the Catholic tradition and in a religious and spiritual view of the meaning of human existence.

Many Australian Catholic educators argue that this is precisely what they are doing; and in many cases they are right: they are offering a genuine alternative. But there problems which vitiate and frustrate the Catholic claim that church schools are different. One is government funding; the Catholic system is heavily dependent on both state and federal governments for recurrent costs and capital expenditure. This has brought state and church educational administrators close together. The government has its own educational priorities, which vary with the prevailing political or, more likely, economic ideology. Given the crass monetarist materialism of virtually all contemporary Australian governments, there is a distinct danger that this ideology will invade the church's educational ministry through state-imposed priorities. In this way the government could subsume the church's agenda. The intimate connection between the various Catholic school systems and the agencies of government funding can easily involve a subtle form of seduction. Political scientist Michael Hogan pointed this out clearly some years ago:

No Australian school is completely free from government interference . . . When governments provide funding assistence to schools, inevitably there are strings attached . . . Sometimes the extent of the conditions attached to funding is so pervasive that there seems to be a threat to the very independence of the independent schools. [41]

---

41. Hogan, Michael, *Public vs Private Schools: Funding and Directions in Australia*, Penguin Books, Ringwood, 1984, p. 112.

Hogan explicitly warns the church about the danger of consorting with government bureaucracies. He might also have mentioned the danger of employing in the Catholic system lay educational administrators who were former government bureaucrats. A number of these have come to the church from the public service or from state education departments. Bureaucrats, like leopards, do not change their spots, no matter what system they work in!

The key prophetic challenge that faces the Catholic church and its school system in contemporary Australia is to assert clearly that Catholic schools are not just educational institutions but are part of the ministry of the Catholic church and that this ministry has a different set of priorities from government educational authorities. It is impossible for Catholic schools to offer a genuinely alternative education if they are no more than institutions whose standards and priorities are determined either directly or indirectly by the state and federal governments. The church must spell out unequivocally that its schools are a major aspect of its ministry, and as such, must be independent of *all* government control.

Several consequences follow. The nexus with governmental educational policies and priorities must be broken, except for funding. The church has to demand that the state fund a uniquely Catholic approach to education. If specific political parties are unwilling to do this, appropriate pressure will have to be brought to bear on them. The church must strive to eliminate government control over what is actually done in its education. I am not arguing that the government has no right to check that its funding is not wasted or misdirected; but it is at this point that government supervision should cease. The church needs to strike out on an entirely new and creative approach to education. In fact, it needs to begin to redefine what education means in this country. The Catholic church has an unrivalled educational track record and now educates a quarter of Australia's children; it is quite capable of determining its own priorities. It should be able to set its own educational standards, the goals it wishes to achieve and the methodologies to be used in the process. The sole role of government will be to fund.

No doubt I will be accused of being 'impractical'. The argument will be that government will not fund something over which it has no control. The Catholic church will thus have to establish a new principle — something it has not tried because of a lack of an articulated educational philosophy in the church, a lack of Catholic leadership and the fact that too many Catholic school administrators at the top level are too close to their government counterparts. Educational administrators feel that if the church were to strike out on the basis of its own philosophy, this would upset the 'system' far too much, and too many jobs depend on the maintenance of the status quo. But it is precisely this type of upset that the church should be attempting. The government cannot ignore the Catholic system in Australia; it is too big. Public political pressure has to be brought to bear, so that the church can attain some distance from the government. If there was a sense of vision for Catholic education and its leaders were willing to bite the bullet, rather than going cap in hand to government, a new approach could begin to emerge, independent of the narrow ideologies and priorities of governments and their short-term vision.

Two catchcries of virtually all recent Australian governments are 'privatization' and 'user pays'. The user of Catholic schools has already paid through taxation. This principle has been clearly established and admitted by the major political parties. So let governments live up to these catchcries by funding the 'ecclesiastical private sector' to do what it has been doing more or less successfully for over 1000 years. The Catholic church, after all, invented education in western culture. The medieval universities emerged in almost every case from the schools of theology. In the 16th and 17th centuries, religious orders like the Jesuits, Mary Ward's Institute of the Blessed Virgin Mary (the Loreto order) and the De la Salle Brothers pioneered modern education. The contemporary Australian model of education is a very recent one, an invention of late 19th century secularism. There is no reason why the Catholic church should lack confidence in demanding that it set its own priorities in education. It has a lot more experience in the education business than the state.

## Ministry of the Catholic school

In its defining of meaning of 'alternative' the church needs to reassert clearly that Catholic schools are not just educational institutions but a form of its ministry. Ministry is a specifically *theological* word whose meaning is clearly linked to baptism and to faith. This faith is acted out by each Catholic Christian through the actualisation of their particular 'charism' or gift of the Holy Spirit. One of the most frequently mentioned gifts in the New Testament is that of 'teaching'. The church needs to be absolutely clear about this: Christian education and formation is a ministry that finds its origin in faith and is meant to lead to faith. Faith, therefore, must be a characteristic of those who exercise a ministry in Catholic schools, and a characteristic of that ministry itself. I am not suggesting any form of anti-intellectual fundamentalism here. Catholic teachers must be formed and trained in both a coherent theological belief, as well as in the particular aspect of human culture in which they specialise. I have no desire to drive a wedge between faith and intelligent education. I am simply asserting that Catholic schools are *Catholic* schools, and part of the church's ministry, and, as such, must be subsumed to the priorities of the Catholic church, not those of the government or any other societal institution nor the priorities of the individual teacher.

I am not saying that Catholic schools should become strongholds of sectarianism. When I use the word 'Catholic', I am referring to its original Greek meaning, 'universal'. The Catholic church is the 'universal' church, the church that is inclusive rather than exclusive, open rather than closed. It is unfortunate that the word 'Catholic' has been defined by some in a very narrow, sectarian sense; this book should have already made clear that I understand the Catholic church in broad, generous terms. [42] Because a school is Catholic does not mean that it is not ecumenical, that other people of good will are excluded. On a national average, up to 10 % of students in Catholic schools are not Catholic. There is no reason why the Catholic church should not be enriched by the Orthodox, Anglican and Protestant traditions, or, for that matter, by the wider ecumenism of the great religions. But my emphasis here is the Catholicity of the Catholic school. It is this that defines the nature of the school and creates its ethos.

This wide and generous definition of Catholicity needs to be kept in mind as I explain a third element of what I mean by alternative. Catholic schools are about Catholic Christianity. They are not just 'religious' schools, in contrast to those of the state which are 'secular' schools. Nor are Catholic schools just generalised 'Christian' schools. They do not exist primarily to educate the children of those who prefer some generic form of religious or moral education. They are not there primarily to assist Muslims, or Protestants, or whoever else wants some form of spiritual or moral education. They are quite specifically schools that exist to teach Catholic Christianity. Central to their ethos and formation then will be Christ, the bible, the Catholic tradition, Catholic worship and a serious commitment to

---

42. See p. 52.

Christian living. At the core of the school's teaching will be the assertion of the essentially spiritual meaning of human existence as understood not just by the generic Christian tradition, but specifically by the Catholic tradition. It will also involve the articulation of an alternative life-style that excludes the crass materialism that characterises so much of Australian secularism. All of this will necessitate the reassertion of a strong Catholic identity.

Further, for Catholic schools to be a genuine alternative, the ethos of Catholic Christianity should permeate the entire syllabus. Again, I do not mean some form of anti-intellectual fundamentalism by which the school teaches 'Catholic science', or 'Catholic history', or 'Catholic whatever'. Catholicism is one of the great historical and contemporary attempts to understand the meaning of human existence, with a coherent and sophisticated philosophy and apologetic. It is this understanding which provides the context for the entire teaching and educational work of the school. It should be the integrating and cohesive force that draws the whole educational effort of the school together. The compartmentalisation of different branches of knowledge is often criticised today. I am suggesting that in the Catholic school, Catholicism can act as the cohesive element that draws the whole educational effort together.

The sad thing is that very few Catholic teachers have the knowledge and religious sophistication to be able to achieve this. School authorities often complain that even school chaplains are unable to provide this comprehensive vision. The cause, which is only now beginning to be tackled, is the almost complete lack of serious tertiary theological education among teachers in Catholic schools.

It must be said that teachers are not entirely to blame for this. The Catholic church in Australia has not made any serious attempt at developing apologetics since the 1930s and 1940s, when Archbishop Sheehan's *Apologetics and Christian Doctrine*, MSC Father Leslie Rumble's *Radio Replies*, Jesuit Father H.A. Johnston's *Apologetics and Christian Doctrine* and the the extraordinary range and circulation of the Australian Catholic Truth Society pamphlets tried to set out a Catholic apologetic. [43] Edmund Campion points out that from the foundation of the ACTS in 1904 it published 1200 titles with a total print run of 13 million. The hope for the future is

---

43. For an historical outline of 20th century apologetics in Australia see Campion, op. cit., pp. 128–137.

that teachers will begin to use modern apologetic classics, such as Hans Küng's *On Being a Christian*. [44]

## Social justice

A fifth aspect of being alternative: one of the most promising things about contemporary Australian Catholic schools is their increasing emphasis on the importance of social justice. For a school to be 'Catholic' in any sense of the word, the church's traditional teaching on social justice and moral integrity must be a core part of the curriculum. Documents such as the bishop's draft statement on wealth distribution in Australia, *Common Wealth and Common Good* could easily become part of a high school syllabus. This emphasis can reach into the primary school too, especially through a stress on what is referred to as the 'common good'. This refers to the fact that human persons achieve their potential as members of society by respecting the needs and rights of others. The emphasis is shifted away from the individual to the importance of the community and to the necessity of working to build up the community.

While the social justice emphasis is already being used extensively in Catholic schools, it can be a mistake to teach the right theory, while being blind to what is actually happening among students, families and teachers in the school itself. For instance, the application of the principles of social justice must create really difficult questions for Catholic schools that cater for the needs of the wealthy and professional élites. At the risk of offending some people, I need to be very clear here as to which schools I am referring. I am not talking about the systemic Catholic schools, most of which do serve the working class and the middle class. Also, I am not talking about the many middle class private schools still run by religious orders which also attempt to serve ordinary Australians. The schools I have in mind here are the exclusive and expensive private Catholic establishments whose fee-structures exclude the children of all but the wealthy or upper middle class. They are normally members of the exclusive group of schools that are known in some states as 'greater public schools'. For a long time the small number of Catholic schools in this bracket — and they are a small minority — justified their existence on the grounds that through the faith education of the children of the rich, Catholic values gained entree into the powerful,

44. Küng, Hans, *On Being a Christian*, Doubleday, New York, 1976.

decision-making sections of Australian society. Nowadays they would argue that their *raison d'être* is to offer quality education. But the problem is that the corruption wrought by wealth and power pervades and subsumes even the strongest faith, and the best established institutions.

The danger they face, of course, is that the loss of parental support and the old boy/old girl networks, would place them in a very difficult position, because the professional and business successes of their parents and former students provide them not only with material benefits, but with the kudos that keeps the school in the élite league. In my view, there is little justification for the existence of such wealthy Catholic schools, unless they can show evidence that they teach and insist on a comprehensive and complete religious education which includes a strong emphasis on social justice as it applies in the Australian context.

There is a sense in which the so-called 'élite schools' are actually quite reactionary. They are about the business of reinforcing the status quo — a far cry from the role of an élite. Ivan Illich, whom many pragmatic Australian educators rejected because they were unable to comprehend the broad range of his thought, has argued that the school has become the key initiation ritual through which the young have to pass in order to imbibe the established values, mores and processes of modern society.

The school system today performs the threefold function common to powerful churches throughout history. It is simultaneously the repository of society's myth, the institutionalisation of that myth's contradictions, and the locus of the rituals which reproduce and veil the disparities between myth and reality. [45]

The opportunity offered to the Catholic school is precisely to challenge prevailing Australian myths and values with the alternative value system of Christianity. It is within the context of challenging societal myths and values that the word 'élitism' can be redefined to refer to genuine creativity. This is where I think the Catholic church can be most educationally creative.

In common speech 'the élite of society' refers to the power-brokers, to the wealthy and to the 'respected' professions (such as medicine and the law), that is, membership of the 'élite' is based on earning

---

45. Illich, Ivan, *Deschooling Society*, Penguin Books, Harmondsworth, 1973, p. 43.

capacity; in this context it has little or nothing to do with genuine ability. The influential position of Catholic education in Australia means that it can transform its schools into academies that train young people to move beyond the prevailing myths of Australian culture to push out the boundaries, to explore new alternatives, to develop other possibilities. At the core of this process will be the confrontation between 'Australianism' and Catholicism. In other words, the word 'élite' needs to be redefined to refer to schools that are genuinely creative.

One of the areas where this might be explored is in the area of Catholic schools refusing to educate students for the contemporary rat-race. In the present economic climate, many students are destined to achieve little more than a deep sense of failure as they compete for scarce jobs. Perhaps the time has come to develop in them different expectations of life and society. In this the Catholic education system is uniquely placed to offer alternatives. Christian faith calls for a co-operative and less competitive approach to life. It is also about trusting God and not placing all security in material 'nest eggs'. It is obvious that the advertising agency and the insurance company which devised the dreadful television ad about 'nest eggs' knew nothing of the gospel text about the lilies of the field and the birds of the air. As Jesus says: 'Do not be anxious saying "What shall we eat?" or "What shall we drink?" or "What shall we wear?" Your heavenly Father knows that you need them all. Seek first his kingdom and his righteousness, and all these things shall be yours' (Matthew 6:25–33). It is precisely by beginning to confront students — and their parents — with some of these radical gospel alternatives that the Catholic church can begin to become a serious force for creative change in Australian society. This is one of the real 'alternatives' that the Catholic school ought to take up: educating young people to question society's norms and values and helping them to imagine other options and new and creative ways of doing things.

## The role of teachers

A sixth consequence of the Catholic school offering a genuine alternative affects teachers. If Catholic schools are a ministerial expression of the church, then what is the role of teachers? The profession of teaching in Australia has a long history that is rooted in the state education system. It also has a long tradition of industrial unionism. The tradition in the Catholic schools has been different. As we have seen, until the 1970s Catholic schools were staffed largely by members of religious orders whose lives, at least theoretically, were devoted totally to their ministry. It was the raison d'être of their being religious; they had no family or social life to distract them. Now that that pattern has changed, what new expectations will be placed on lay teachers? How much time will they be expected to give to this ministry? How will they balance family and social commitments with their ministry in the school? Obviously, this will be worked out in different ways by different people; it is impossible to give a normative answer, except that it will involve a lot of self-giving and a willingness to go beyond the minimum and even sometimes as far as the maximum.

This leads to inevitable tension between what Kevin Treston has called the 'vocational' and 'industrial' models of teaching. The industrial model refers to terms and conditions worked out by the union between employer and employees. Treston has used the word 'vocational' to refer to the extra commitment required of the teacher in the Catholic school. I think, however, that the use of the word 'vocational' in this way is inaccurate. There is a sense in which *all* forms of teaching are vocational. It is a serving profession that requires a degree of commitment to students, to their families and, in the broadest sense, to the betterment of humankind. The church's ministry also asks for this, but it asks for this commitment from within a different context. Ministry in the church is an explicitly theological call to work within the context of faith. Thus the tension for the Catholic lay teacher is more likely to be between the 'industrial' and the 'ministerial' models. I will now try to sort out what I think are some of the consequences of this tension.

The Catholic church should not be opposed to, or in any way afraid of trade unions. As probably one of the few priests in Australia in a union (as an employee of the ABC I belong to the Public Sector Union), I am a firm supporter of organised unionism. In a time of so-called 'economic rationalism', unions are a needed guarantee of social justice and equity for the ordinary worker. Even the *Code of Canon Law* seems to agree that workers need protected rights:

They [lay workers in the church] have a right to a decent remuneration suited to their condition; by such remuneration they should be able to provide decently for their own needs and for those of their family with due regard for the prescriptions of civil law; they likewise have a right that their pension, social security and health benefits be duly provided. [46]

---

46. *Code of Canon Law*, Latin-English Edition, Canon Law Society of America, Washington, DC, 1983, Canon 231, 2.

As an employer, the church should not attempt to use ministry as a way of exploiting its workforce and it should encourage those whom it employs to belong to a union.

There is another important premise that the church must accept in the implementation of its educational ministry: stated bluntly, it is that there is no place in a Catholic school for teachers who have no sympathy for religion. Certainly there is no place for those who are opposed to Christian faith and spirituality, nor for those who do not experience, or at least accept and comprehend the possibility of such conviction. Those who consider that faith is meaningless or even nonsense, cannot participate in a ministry that is motivated fundamentally by faith. Before any person is employed in a Catholic school they must be prepared to honestly state their position on religion and faith. This will need to be done with sensitivity, and such a stance could create problems with equal opportunity legislation in some states, but it seems to me dishonest not to tell prospective teachers clearly what the Catholic school is all about, and to seek from them some form of commitment to that ideal. This is another one of those situations where the Catholic church will have to be strong enough to make a stand against state legislation that does not take into account its particular needs.

Most dioceses in Australia are already moving toward an accreditation process that will require all teachers in Catholic schools to undergo in-service training and formation in the theological understanding that underpins Catholic education. The Catholic Education Commission of Victoria, for instance, emphasises that 'all teachers in the Catholic school are expected to share in its distinctive concerns'. [47] However, Catholic educational administrators still seem very tentative in this area. They need to become more honest and definite. The church needs to become crystal clear on the fact that it will only employ people in its schools who share its understanding of the ministry of teaching. This is not meant to be anti-ecumenical or to propose that only Catholics can be employed in the church's schools, but it is to demand that those teaching in Catholic schools have sympathy with the ultimate purpose of Catholic education. The Catholic Education Commission of Victoria states clearly what this purpose is: 'all members of the school derive their understanding of its educational purpose and vision from the person

---

47. Catholic Education Commission of Victoria, *Circular to Principals of Catholic Schools in Victoria*, 5 February 1988, p. 4.

of Jesus Christ who continues to be present in and through his church'. Much has already been achieved in this regard in Catholic primary schools. Most of the teachers at this level are Catholic and there is clear evidence that the vast majority of them are committed to religious education and to what Brother Marcellin Flynn has called the 'informal curriculum', that is the maintenance of the Catholic ethos of the school. Part of the reason for this high level of commitment from primary teachers seems to be that they all have to teach religious education directly, because they work consistently with one class and teach the children all subjects.

In secondary schools the situation is different. Only a minority of teachers are directly concerned with religious education and many do not have any advanced training in theology. Secondary teachers are trained in specific subject disciplines and are usually employed because of their expertise in that subject. The Catholic system has moved, certainly, beyond the period of the late 1970s when you could meet teachers in Catholic secondary schools who, either overtly or subconsciously subverted the maintenance of a Catholic ethos by their anti-religious or anti-Catholic attitudes. However, while there has been a major change for the better in the attitude of secondary teachers, the question of the 'informal curriculum' is still an important one. Marcellin Flynn quotes the American educationalist, Otto Kranshaar, to describe what he means by this:

The evidence is substantial [that] . . . the religious setting of the school does make an important difference. It is not what the teachers say, but what they do; not the content of religious and moral instruction, but the life within the school. The atmosphere, the outlook and expectations, the dedication of the staff . . . in most schools in which religion is a vital influence, are unmistakable. [48]

To share in this vision, a teacher does not necessarily have to be a Catholic, but at least he or she must be a person who is religiously sympathetic.

Having said this, it is important, however, to focus attention on the key people in establishing and maintaining the Catholic ethos of the school. These are the principal, the deputy principal, the religious education co-ordinator and the major subject heads. The Catholic community has a right to demand that Catholic educational authorities do not compromise on any of these positions. These people have to be committed Catholics with professional ministerial and

---

48. Quoted in Flynn, Marcellin, op. cit., 1979, p. 148.

theological training. Throughout this book I have emphasised that ministry is not for good-intentioned but ill-informed amateurs and this is especially true in the case of key people in a Catholic school. There has to be a continuing insistence on theological training for secondary teachers in key positions in the schools. These are the people who set the tone for the school's ministry.

In the light of this it is interesting to note that the Victorian Catholic Education Commission's accreditation document eschews the use of the word 'ministry'; it prefers 'vocation'. One reason for this is obvious: it wants to side-step the debate about whether the word 'ministry' should only be applied to the ordained ministry. The official Roman view is to restrict the word 'ministry' to the hierarchical ministry, and it is obvious that the Victorian Catholic Education Commission would not want to offend the sensitivities of the Vatican. But perhaps, more importantly, the commission is making a subtle, but nevertheless real distinction, between accreditation to teach in a Catholic school, which it probably correctly calls a 'vocation', and explicitly teaching religious education which is a further step. The commission does not call this a 'ministry', although the whole weight of contemporary theological opinion would. I would argue that since all teachers contribute to the ethos of the school, they are all involved in the ministry of the school.

Before moving on from teachers I want to mention the fear that many Catholic teachers have of participating in the religious education programme themselves. Many feel inadequate through lack of theological formation, or experience a crisis of confidence in themselves because of what they subjectively experience as a lack of faith. It is unfortunate that good people are frightened away from a ministry in which they might have much to offer to the students. Over the last decade there has been a serious attempt by Catholic educational authorities to offer in-service training in both religious education and theology. This will also need to emphasise personal spiritual and faith formation for teachers so that they may gradually gain the confidence to understand that their seeming lack of faith is really a challenge to move toward trusting God at a deeper level.

Sometimes this feeling of inadequacy and unworthiness arises from what could be called a particular 'life dilemma' of the teacher. For instance, some still feel that because they are using contraception, or are involved in marital infidelity, they are not worthy to teach religion. The simple answer to this is that if the ministry could only be carried out by saints or the morally perfect, very little would be achieved. This, of course, can become more difficult when relational situations become public — for instance, when a teacher is living in a de facto situation, or has been divorced and has then married again outside the church, or when a man or woman is living in a homosexual situation. Generally speaking, the Australian Catholic church has dealt with these examples rather well. While there have been a few unfortunate and public disputations, in most cases, as long as the person involved does not set out to mount a public confrontation, situations are dealt with with a considerable degree of compassion and tolerance. Things can get difficult, of course, when crusading parents or priests interfere. There is a simple answer to such people: 'Let the one without sin cast the first stone' (John 8:7). As I said above about priests: it is the person who has sinned and who, like Saint Peter, the first pope, has betrayed their own deepest ideals, who can be equipped to teach and minister to others. If we excluded sinners from ministry, there would be no ministry!

## Schools are for students

I want to turn now to the students, the products of Catholic schools. Here I need to make a another confession: I am not only a failed parish priest, I am also a failed Catholic teacher. A year of primary teaching in the early 1960s and two of secondary teaching cured me of education forever — and no doubt cured the students as well, given that my nickname as a teacher at Downlands College in Toowoomba was 'Heinrich' (as in Himmler!) I still wince when I occasionally meet former students, remembering that I often lived up to my nickname.

Before talking about students in particular, I will begin with the now tediously repeated claim by reactionaries that there is no 'content' in religious education in Catholic schools, that students do not know doctrine, have not been taught to go to Mass, do not know their prayers, or the ten commandments, or whatever! While some of these criticisms were valid a decade or so ago, it is no longer true that primary and secondary religious education in Australia lacks content. Religious education programmes have adopted more educationally sound teaching strategies and learning processes, and students are no longer involved in rote learning. Living faith is not primarily belief in a series of propositions; it precedes propositional faith as an existential reality that is primarily expressed through life itself. Certainly secondary school students need to learn something of the church's theological approach and to be provided with a sound apologetic for belief. But there is absolutely no reason for students to be discussing abstract theological concepts in primary school.

However, recently an attempt has been made to put forward a 'content oriented' draft curriculum for primary schools in Wagga diocese, entitled *To Live in Christ*. The proposed curriculum was viewed by many, both inside and outside the diocese, as an attempt to resurrect a pre-conciliar approach to primary religious education with a heavy emphasis on 'theological' content. But even as an attempt to restore content the draft curriculum was ill-conceived. The effectiveness and potency of the old 'Penny Catechism' was that it was a succinct expression of a whole religious culture and a way of life that was inculcated by the Catholic school, the traditional parish and the Catholic sub-culture.

Needless to say reactionaries strongly defended both the curriculum and Bishop Brennan, bishop of the diocese. In the parish of All Saints, Tumbarumba, for instance, the parish notice sheet told parishioners:

There has been a lot of malicious comment in local and major city newspapers concerning this curriculum. Most of the comment has been more a direct personal attack on the Bishop and his authority and office. The time has come for us to get off the fence and declare ourselves for or against the Bishop and his right to exercise this office in a correct and proper way in common with the Pope (Parish Bulletin, 10 March 1991).

Parishioners were asked to sign a declaration of support for Bishop Brennan and the curriculum. Predictably *AD 2000* weighed in with an article entitled 'Dissident priests and teachers gang up on Bishop Brennan (April 1991). The issue of the proposed curriculum was widely discussed in the media, especially by Albury's *Border Morning Mail* and also on ABC radio. The curriculum was debated at public meetings in Wagga, Albury and Narrandera. Seventeen priests of the diocese signed a letter which said that 'So strong are our convictions about [the curriculum] . . . that we could not allow such guidelines to be introduced into any school for which we have responsibility'.

Early in 1991 Bishop Brennan set up a review committee from outside the diocese to report to him on *To Live in Christ* and to propose a future management plan . . . to enable a Religious Education Curriculum for Primary Schools to be introduced (30 June 1991). The *Report* of the review Committee says, 'There was [an] almost unanimous request from most parents that something of substance be provided as a basis for religious education classes' (*Report*, p. 25).

But the *Report* was critical of *To Live in Christ* as a response to the need for something substantial. It criticized the draft curriculum for selective and incomplete use of the Vatican document *Catechesi tradendae* (*Report*, p. 19); many of the theological concepts used in the draft curriculum are described as 'too difficult for primary school children' (*Report*, p. 19); it asserts that 'the full spectrum of scholarship available to professional Religious Educators . . . is deficient'. (*Report*, p. 20); scripture texts are only used as 'a proof to support doctrinal statements' (*Report*, p. 22); 'the Core Content section is deficient in the area of ecclesiology in that it reveals an incomplete understanding of the Church' (*Report*, p. 23). These are fairly serious criticisms and, in fact, describe the usual limitations built into most reactionary attempts to drag the church back to the past. It can only be hoped that Bishop Brennan will follow the committee's advice and bring Wagga diocese back to the mainstream of catechetical development in the Australian Church.

To return to the mainstream: recently, the Sydney Archdiocesan Schools Board surveyed clergy, parents and teachers on the topic 'What do you expect of a religiously educated young Catholic?' The results of the survey showed that the 'religiously educated Catholic' would be expected by the respondents:

- to know and love Jesus Christ
- to care for the wants and needs of others
- to have a conviction that they must take responsibility for their actions
- to accept themselves
- to accept that God became human in Jesus
- to be challenged to grow as a person
- to have a conviction that humans are free to make decisions
- to be able to make close friends
- to be proud to be a Catholic
- to have strong moral values. [49]

The list is obviously a hotchpotch of expectations from various points of view within the church and, as the survey itself points out, attitudes 'are strongly related to the age of the respondents'. It is the comments on the survey, rather than the survey itself, that I found most useful.

Firstly, the survey showed up real problems with the language and religious rhetoric used by religious education professionals. It indicates that we are at an intermediate stage between the religious language

---

49. Sydney Archdiocesan Catholic Schools Board, *The Religiously Educated Young Catholic*, Survey, 1990. The document was prepared by Graham English of the Sydney CEO.

of thirty years ago and Vatican II rhetoric which 'seems to have little meaning for most people'. This tends to be reflected also in a lack of clarity on people's part about Catholic identity and in expectations that Catholics have about their schools.

The evidence in this study is that adults younger than fifty . . . will not in the future place the same strong expectations on Catholic schools that former generations did. This may lead to a greater emphasis on family and parish catechesis by practising Catholics with consequent higher expectations of parish and diocesan adult education opportunities and family support programs.

This same lack of clarity is reflected in the definition of what is meant by 'good behaviour'. Repondents clearly want Catholic students to have moral standards, but they are not sure what these standards are. The clearest thing that seems to emerge from the survey is that 'Catholic people associated with our schools want students to leave them imbued with the Catholic tradition'. Secondly, the survey showed that Catholics value the sacraments and the bible very highly. I think that if Catholic schools inculcate a sense of the Catholic tradition and convey a sense of the centrality of the sacraments and the bible, they have succeeded admirably.

This leads directly to the question: are the schools, especially the secondary schools, achieving these expectations? What is the effect of religious education and Catholic schooling on students at school and what happens after they leave? Does it make any difference that they have been to a Catholic school? It is difficult to answer the second of these questions because it involves serious longitudinal studies of former students. One person who has attempted to do so is Jesuit Father Noel Ryan who did a major study of students who matriculated through Victorian Catholic schools in the decade 1950 to 1960. He found a high faith retention rate, with 85% of students still practising as Catholics ten years after they left school. [50] This is a remarkable achievement, but it needs to be remembered that forty years ago only the best and most committed students stayed on to reach the matriculation class. It was also the period of stability just before the student revolutions of the 1960s and 1970s in education, and the radical changes in the church prompted by Vatican II. A level of stability is to be expected from people of that generation.

Ryan's student subjects are now people in their 50s and they are

---

50. Quoted by McCarthy, F., in *Catholic Education in Victoria: Yesterday, Today and Tomorrow*, Catholic Education Office of Victoria, no date, p. 47.

the key generation who have bridged the process of change from the pre-Vatican II church to the church of today. We do not know what has happened to these people subsequent to the early 1970s. Their rate of practice has most probably dropped to the general contemporary Catholic level that I outlined earlier. What the Catholic school has probably done for them is to make sure that Catholicism is an integral part of the texture of their lives. Can the same be said of students over the last two decades in Catholic schools? These are the products of Vatican II properly speaking. What has happened to them?

Before I address this question I want to refer to one genuine longitudinal study of the effectivness of Catholic schools. Sociologist Andrew Greeley has been looking at Catholic schools in the United States since 1966. During that time the American church more or less lost the battle to maintain its schools. However, Greeley argues that the surviving schools are still very effective.

The influence of the Catholic school on the religious behaviour of young Catholics is stronger than that of the family of origin. It is only slightly less strong than the family of procreation . . . The old explanation of the success of Catholic schools, that they were merely duplicating the work of the Catholic family, is simply not valid. Catholic schools seem to have their effect on those who attend them not so much through formal religious instruction class but rather through the closeness to the Catholic community that the experience of attending Catholic schools generates. [51]

Greeley argues that his continuing research shows that Catholic schools are not becoming less effective but more so.

It is worth keeping Greeley's research in mind as we approach Australian material. It is fortunate that the Australian church does have a major long-term study of the effectiveness of both religious education and the Catholic school. The work has been done by Dr. Marcellin Flynn, a Marist brother, now teaching at the Strathfield campus of Australian Catholic University. [52] Much of Flynn's material parallels that of Greeley. One difference here is that, because of the strong leadership provided at the time of the granting of state aid, Catholic schools in Australia, unlike their United States counterparts, have not only survived but actually prospered. Marcellin Flynn has just completed another major survey of year 12 Catholic students and he assured me that the new material confirms what

---


51. Greeley, Andrew, *The Catholic Myth: The Behaviour and Beliefs of American Catholics*, Scribners, New York, 1990.
52. Flynn, Marcellin, *The Effectiveness of Catholic Schools: A Ten-Year Study of Year 12 Students in Catholic High Schools*, Saint Paul Publications, Homebush, 1985.

he said in his 1985 book *The Effectiveness of Catholic Schools.* [53]

What do Flynn's findings tell us? One of the strongest conclusions concerns the climate and ethos of Catholic schools:

One of the most distinctive features of an effective Catholic school is its outstanding social climate which gives it a special ethos or spirit. This climate has a religious as well as an educational character and is generated by an intensely relational environment in which persons are respected and ultimate questions such as life, love, death, faith and God are confronted. [54]

There is increasing evidence that students are satisfied with Catholic schools for they provide a sense of community, bonding and belonging. It is clear that there is much less alienation in Catholic schools and very little teenage suicide, an increasing problem in some schools. Catholic students clearly appreciate the relational atmosphere created in most Catholic schools. The schools also clearly provide a high level of pastoral care for the students. There has been, however, a continuing decline in the religious values and practice of students. Regular attendance at mass has dropped from sixty-nine per cent in 1972 to fifty-five per cent in 1982 to thirty-four per cent in 1990. A similar decline can be seen in the practice of the sacrament of reconciliation. This parallels the decline that can be traced among adults. Flynn's 1990 research shows that while there has been a drop in the influence of the home, the family still has an important influence on Mass attendance and sacramental practice.

One of the great success areas has been school retreats:

A major change observed since 1972 concerned the importance of religious experiences such as school retreats, Christian living camps and weekends. After the family, these retreat type experiences in 1982 were the single most important religious influence in the lives of many students. [55]

Among young people — as among older people — there is clearly a longing for spirituality and for deeper prayer. Flynn also comments that

Evidence was found for a movement towards an autonomous personally-owned faith in many students. It is not yet personal faith in Jesus, but the re-emergence of personal faith. [56]

---

53. The conclusions of Flynn's 1990 survey are available now and will soon be published.
54. Flynn, op. cit., p. 342.
55. ibid., p. 347.
56. ibid., p. 349.

However, there was evidence of alienation from the church and its teaching authority. 'This alienation stands in striking contrast to their love for Jesus, their perception of God as a loving, caring Father and their warm regard for Catholic schools'. [57] Flynn correctly points out that this is to be expected because late adolescents are deeply involved in the process of separation from the family matrix and value system, and they are often also in the process of rejecting an inadequate understanding of the church.

The data that Marcellin Flynn has produced certainly seems to indicate that Australian Catholic secondary schools are far more successful in inculcating a religious perspective than most people expect. Students are certainly not alone in their crisis of credibility with the institutional church; they share this with many adults. Given that so much of Flynn's data tallies with that of Greeley in the US, it is abundantly clear that the continuance of Catholic education, at both primary and secondary levels, is essential for the healthy future of the Australian church.

## Education in the future

What then ought be the future emphases for Catholic education — and please note, I said 'education' rather than 'schools'? As I mentioned at the beginning of this section, schools are an invention of the 19th century secular liberal reformers, while the church has been in the business of education for at least a millenium. It ought value this wealth of experience. I also suggested that the church must discover what an 'alternative' education means, for it is in a unique position to offer whole new ways of educating. It is not just a matter of teaching new things, but going about education in a different way. Perhaps we need to develop further the real sense of community that already exists in Catholic schools. An increasing emphasis on social justice might imply that students be invited to begin to share in the life of the poor; in other words that they move from the realm of theory to practice. As the centrality of environmental issues increasingly engages the ethical debate, perhaps Catholic students could be involved in a 'hands-on' experience by caring for an environmentally degraded area close to their school.

Catholic education also has to see itself as part of an increasingly multiform church in which there are many valid ways of being Catholic. This has always been true, but has been obscured in the life of the church by a post-reformation emphasis on the identification of unity with uniformity. One of the dangers inherent in a

---

57. ibid., p. 346.

bureaucratic centralising of the Catholic school system is that it will prevent the development of a legitimate pluralism within that system.Ivan Illich's warning about the school as an instrument of inculcating the prevailing social ideology should be taken to heart. The Catholic school should deliberately set out to break down class and racial divisions. Perhaps Catholic education systems in some of the major metropolitan areas need to begin some form of bussing — and equalisation of fees — to achieve a reasonable mix of students. I was living in Boston at the time of the school bussing crisis in the late 1970s. Black children from the ghettoes were bussed, on court orders, to wealthy white neighbourhoods, so that there could be some equality of opportunity in education. This reflected, of course, the pecularly local nature of educational financing and administration in the United States, and it was a brave attempt at a form of social equalising. Certainly, a church which espouses a social justice ethic cannot allow the enshrinement of social inequality.

Whatever the future, there is no doubt that much has been achieved by Catholic education in Australia and there is no doubt also about its continuing importance in Australian society, with close to 30 per cent of all children now in Catholic schools. At present there is something of a crisis of confidence among Catholic educators, but this should be seen not as chance to retreat to a comfortable sub-culture but as an opportunity to move onward creatively. Both the Catholic community and a healthy pluralism within Australian society require this.

# *Seminaries and universities*

## Catholic seminaries

Seminaries are a comparatively modern invention in the Catholic church. At the final session of the Council of Trent (1562–1563), as part of its practical attempt to reform the priesthood, the Council decreed that priests ought to be given a formal education and training in seminaries. But it was really not until mid-17th century France that something practical was established on a wide scale. The pioneer of modern seminary education was the French priest, Jean-Jacques Olier (1608–1657). He used his parish of St Sulpice in Paris as a base for training student priests. The congregation he founded, the Sulpicians, became pioneers of seminary education in France. Saint Vincent de Paul, the most well-known saint of the 17th century, founded the Congregation of the Mission, or the Vincentians as they are called in Australia. His congregation was also involved in pioneering seminary training in 17th century France.

The first Irish seminary was Saint Patrick's College, Maynooth, established in 1795 in a village just outside Dublin, with a financial grant from the British government. (Previously there had been Irish — and English — seminaries on the continent from the beginning of the 17th century to the French Revolution). Maynooth was based on the French seminary model and even had a number of emigré French priests on the pioneer staff, exiles from the French Revolution. Subsequent seminary foundations in Ireland followed the Maynooth model, but without the intellectual emphasis. The Irish model was brought to Australia — with Roman adaptations — when Saint Patrick's College, Manly, was established by Cardinal Patrick Francis Moran in 1889. The preparatory seminary — Saint Columba's College at Springwood in the Blue Mountains — was begun in 1909. Both were firmly under the control of the Irish — and later Australian — diocesan clergy. Corpus Christi College, the Melbourne seminary, was established in 1923 in the former Chirnside mansion at Werribee, south west of Melbourne. (Corpus Christi Seminary is now situated in the Melbourne suburb of Clayton, after a brief sojourn in the monumental white-elephant seminary of Glen Waverley). In 1922 the Vatican had decreed that Manly was not the 'national seminary' and had encouraged the establishment of other Australian provincial seminaries. This was a sufficient spur for Archbishop Daniel Mannix to establish his own seminary, with different attitudes to life and ministry from those ingrained at Manly.

Mannix entrusted the seminary to the Jesuits. Patrick O'Farrell describes the differences between Werribee and Manly:

In contrast to Manly with its Irish staff and Roman disciplinarian traditions, the Jesuits at Werribee were recruited not only from Ireland, but also from England and America — and Australia. Mannix gave them a free hand, but they were very much in accord with his own view that the clergy needed not only a deep spirituality, but a sound academic training combined with the encouragement of intellectual initiative . . . the Werribee staff did not remain isolated in their seminary, but took an active role in the general Catholic community. [58]

O'Farrell suggests that Werribee priests were more open to 'the

58. O'Farrell, Patrick, *The Catholic Church and Community in Australia: A History*, Thomas Nelson, Melbourne, 1977, p. 365.

Australian environment as the field for an out-going apostolic Catholicism'. Certainly, there was — and still — is a subtle, but discernible difference between the attitudes of the Sydney and Melbourne clergy. But, as O'Farrell also suggests, the difference between the two seminaries can be exaggerated. Both were subject to the narrow orthodoxy imposed by Rome in the years after the so-called 'modernist crisis' (the 1920s to the 1950s). Both were monastic in their approach to formation and spirituality; young men were formed for a life of ministerial involvement in the world as diocesan clergy through a process of isolation from the very milieu in which they would work. The other capital cities were well behind Sydney and Melbourne in the establishment of seminaries: Brisbane's Pius XII Seminary was opened at Banyo in 1941 and in 1942 seminaries were established in Adelaide (St Francis Xavier Seminary, Rostrevor) and Perth.

The clerical religious orders also gradually began to train their candidates in Australia. The pioneer was the first Archbishop of Sydney, John Bede Polding, who set up a seminary in Woolloomooloo as early as 1839. Because his ideal was to establish an 'Abbey-Diocese', he wanted his diocesan priests to be monks. Most of them simply wanted to be diocesan priests and the experiment failed. After the foundation of Manly, Father Pierre Tréand, a French Missionary of the Sacred Heart, was the first to educate Australian priests in Australia for his own congregation. By the 1970s most of the larger clerical orders trained their ordinands in Australia, many of them in Melbourne.

Something of the feeling of the seminaries of the 1940s and 1950s has been vividly described by Thomas Keneally in his early novel *The Place at Whitton* and in the brilliant and funny *Three Cheers for the Paraclete*. [59] Both are obviously set at Manly seminary. The novel *Three Cheers* describes an Australian seminary characterised by intellectual narrowness, oppressive and unimaginative personal formation and sexual rigidity. There is considerable truth in Keneally's dramatic picture. However, as one trained almost a decade later, in the seminary atmosphere between 1956 and 1967, I would have to say that the ethos in my time was a lot less oppressive personally and a lot more stimulating intellectually than that described by Keneally. Admittedly, our lives were very confined; I can remember times in the seminary in Canberra (where I studied philosophy for three years) not even seeing a woman for weeks on end. However,

---

59. First published by Angus and Robertson in Sydney in 1968. Later available as a Penguin book.

I found that this lack of social interaction helped to focus the mind with rare clarity!

I was trained in a religious order and the clerical orders usually had a better community atmosphere than the diocesan seminaries. Also, I did theology during the Second Vatican Council, which was a great advantage. As students, we were able to watch the drama of the Council unfold. One of the blessings of my life was — and still is — a thorough formation in philosophy, theology, biblical studies and church history. A serious attempt was made in my seminary studies to develop a logical and critical sense which has stood me in good stead, especially later in the world of tertiary education in both Australia and the United States, where I discovered that logic and intellectual self-criticism were often lacking.

However, soon after I left the seminary in 1967, things changed radically. From about 1970 onwards, students were given much more freedom, both personally and intellectually. It was the period of instability and change after the Council. Intellectually, there was the struggle to integrate the new vision of the church that emerged from Vatican II, a new understanding of the priesthood and the development of a ministerial spirituality. At this time some terrible misjudgements were made, both in personal and intellectual formation. In my view, the abandonment of philosophy as a coherent discipline in intellectual formation was the worst mistake. As a result, it has become increasingly difficult for students to tackle serious theology. Seminarians today lack the coherent theoretical understanding of the meaning of human existence that flows from a solid study of philosophy. In personal formation the oppressive discipline of the old days was abandoned and an attempt was made to encourage students to develop self-discipline. A more extensive psychological assessment process was introduced, although, despite this, it is clear that questions still need to be asked about the quality of candidates entering the seminary. Despite the criticisms of reactionaries, it is obvious that since the mid-1980s, formation in the seminaries has settled down and a new regimen of priestly formation is widely accepted in Australia. It is true that for a period of ten to fifteen years seminary training was in chaos, but that was to be expected as a a 300-year-old institutional model was adapted to fit new emphases and circumstances.

According to the *1990 Catholic Directory* there were 221 diocesan seminarians and 152 clerical religious seminarians in Australia in that year. As I have already mentioned, this is a very large decrease over numbers in the 1960s; it is certainly not a sufficient replacement rate.

Thus the Australian church will soon be forced to look for other candidates for ordination, including married men, and — eventually — women. However, these figures need to be seen within the context of an increase in the number of seminarians world-wide. According to the *Statistical Yearbook of the Church*, between 1988 and 1989 there was a worldwide increase of 1749 seminarians. The largest number was in Africa, followed by Asia and Latin America. There was a decrease of almost 500 in Europe and fifteen in Oceania. [60]

The major problem facing seminary formation today centres on the church's understanding of the nature of the priesthood. It is difficult to form men for a ministry that is, as I have already mentioned, still in the process of refocusing its priorities. One thing, however, is clear: the priest will continue in the immediate future to be the liturgical and pastoral leader of the local Christian community. Candidates for ordination will need to have a psychological profile that goes with community leadership and they will need to develop a spirituality that will support them in that ministry. A close scrutiny will have to be given to those accepted into the seminary for priestly formation. The problem is that, while a reasonably reliable psychological assessment has been introduced, the number of candidates has dropped so sharply that the temptation is to accept any male who applies and who vaguely meets the basic norms. And psychological assessment brings its own problems: the very process of assessment may exclude creative people as well as problem people. As I have already said, it is clear that the church must allow a wider basis of selection by breaking the nexus between the priesthood and celibacy.

My own impression is — and this is confirmed by checking with a number of experienced priests — that in the last few years, the personal and intellectual quality of the men studying for the priesthood has improved considerably. Even to contemplate the priesthood and celibacy in contemporary church and society requires rare courage and considerable clear-sightedness. However, there is also evidence that a sizable coterie of reactionary types have tried to enter the seminary. It is to the credit of almost all Australian seminaries that they have tried to face the difficult task of making sure that only men adjusted to life and ministry in the contemporary church are ordained.

---

60. *The Statistical Yearbook of the Church* quoted in the *Catholic International*, 2/8(1991), p. 363.

There are still two options for candidates dissatisfied with the mainstream seminaries. They can apply to the schismatic Lefebvrist seminary at Goulburn south of Sydney. This will exile them totally from the Catholic mainstream, but it may well fulfill their apparent nostalgia to live in the 19th century. Another alternative is for them to apply to the newly established Vianney Seminary (named after Saint Jean-Marie Vianney, the Curé of Ars) at Wagga Wagga in the New South Wales Riverina. This is the creation of Bishop William Brennan, who claims that students trained in urban seminaries, such as Manly or Clayton, return ill-adjusted to the rural environment. However, it has been asserted that the real reason for the establishment is suspicion of the post-Vatican II theology taught in the mainstream seminaries.

## Colleges of theology

So far we have looked at seminaries as institutions for the education of priests. Increasingly they have also become places for anyone who wants to study theology at a professional level. This has been one of the most important results of the movement of theology and the seminaries into the mainstream of tertiary education. Until very recently, theology was excluded from Australian universities and confined to the training institutions of the various churches. The only exceptions to this were the small School of Divinity at Sydney University and the Melbourne College of Divinity (MCD), which was established by an Act of the Victorian Parliament in 1910. [61] In the 1970s in Melbourne, where most of the best professional theology in Australia has always been done, the various church training colleges came together under the sponsorship of the MCD. It was a far-sighted ecumenical venture. As Neil Ormerod says:

This brought together Catholic, Uniting Church, Baptist, Churches of Christ and Presbyterian Churches in a united venture. Each institute . . . would retain basic elements of its tradition, but together they would monitor one another's academic standards and promote co-operation. This consortium would then offer a Bachelor of Theology degree, the first purely undergraduate theology degree offered in Australia. [62]

---

61. For a brief history of the development of professionally recognised degrees in theology in Australia see Ormerod, Neil, 'Theology under siege (II): Enter the (Dawkins) dragon', *National Outlook*, 12/6(1990), pp. 8–12.
62. Ormerod, art. cit., p. 8.

This initiative provided the first opportunity for lay people to enter the field of professional theology, especially when courses were offered at night by the constitutent schools of the consortium, such as the United Faculty of Theology in Parkville (itself a combination of Anglican and Uniting Church faculties and the theologate of the Jesuits), Yarra Theological Union in Box Hill (comprising a number of major clerical religious orders) and the Catholic Theological College at Clayton (which includes Corpus Christi Seminary). Catholic Theological College is now entering into a teaching relationship with Monash University.

All of this ensures that theology in Australia is now studied in a truly ecumenical and professional setting. The mainstream churches have turned away from a sectarian approach to theological and ministerial formation, emphasising their interconnectedness rather than their separation. Through personal contact, joint teaching and examinations, and scholarly and student interchange, the long-term ecumenical church of the future is being formed in Melbourne. The pooling of resources has meant that faculties can be strengthened, narrow sub-cultural sectarianism eliminated and genuine tertiary standards of education maintained. The formation of the Joint Theological Library at Ormond College in Melbourne, for instance, has given Australia its best theological collection.

During the 1980s similar theological amalgamations developed in the other states. These were the Sydney College of Divinity (SCD), the Brisbane College of Theology and the Adelaide College of Divinity. Adelaide grants its degree in association with Flinders University. For a brief period in the 1960s and 1970s St Patrick's College, Manly, had the status of a Pontifical Faculty; that is, it could grant degrees up to the doctorate level which were approved and recognised by Rome. This ceased when Manly entered the SCD. Either initially unwilling to enter the SCD, or having withdrawn after entry, Evangelical Christians (mainly low church Anglicans) have developed the Australian College of Theology into a recognised degree-giving institution. However, with the exception of the MCD (protected by its own Act of Parliament) and the Adelaide College of Divinity (through its association with Flinders University), the other consortia are threatened by the Dawkins reorganisation of higher education. Because of their peculiar status, they could easily find themselves isolated and outside the national system, unable to grant degrees. Their only apparent hope is to persuade one or other of the universities in Sydney and Brisbane to recognise their degrees.

And, as Neil Ormerod says, the other 'wild card' in tertiary theological education, for Catholics especially, is the formation of the Australian Catholic University from the former Catholic colleges of advanced education. To this point there is little apparent likelihood that any of the Catholic faculties of theology will join the Catholic University. It would mean divorcing themselves from the mainstream ecumenical context in which serious theology is now largely studied. In Melbourne especially, where the status of the Catholic faculties is protected through the MCD Act, they are unlikely to join an ecclesiastically-sponsored institution. At one stage, it was thought that the Catholic Institute of Sydney (that is the teaching faculty of St Patrick's College, Manly) might unite its excellent theology faculty with the Australian Catholic University. But it seems that this union is not to go ahead. However, invitations have been issued to other Catholic faculties. Neil Ormerod says:

Approaches have been made to the Catholic members of the various consortia to simply become the theology faculty of the new university. While this may solve the problem of degree accreditation for the institute involved, it would also threaten existing consortia with at best a weakening and at worst a dissolution. Many Catholic theology institutes are making it known that their prior commitment is to their local consortium and they are not interested in such a move. [63]

The approaches of Australian Catholic University to Catholic faculties is an unintended but nevertheless real threat to Catholic participation in theology in a genuinely ecumenical setting.

## Issues for the future

Where do the seminaries go from here? In my view a number of issues need to be considered: Firstly, they need to consolidate what has already been achieved, especially in terms of their intellectual formation. The seminaries need to maintain contact with mainstream tertiary education through institutions such as the theological consortia, the MCD and the universities. In this way, essential educational standards will be maintained. The type of society into which students will be moving as priests will increasingly be a 'knowledge culture' that places a priority on high educational standards. The Australian church should not be seduced into believing that all that it needs are 'pastoral' priests, ordained despite their lack

---

63. Ormerod, art. cit., p. 12.

of educational formation. It is a false dichotomy to place educational standards in opposition to ministerial formation. Ministry today requires serious professional theology. The contemporary approach to pastoral theology implies an ability to understand what is going on in our world and to perceive the direction of the culture and the stresses and strains operative within it. It is this reality that provides the context for the pastoral care of individuals and communities.

I am not suggesting that the Australian church needs a bunch of European intellectuals. This has never been part of the Australian Catholic tradition; the clergy here have a history of being close to reality. But, as we have already seen, our country has been dragged down and our cultural life distorted and destroyed by successive federal and state governments made up of 'practical men' — and they are mainly *men* — whose viewpoints are so circumscribed that they know nothing of the big picture. The danger is that narrow-minded, uneducated, 'pastoral' priests, will be as circumscribed and destructively short-term in their views as are so many of Australia's politicians. The church cannot afford to retreat into the type of anti-intellectualism that offers no understanding of the relationship of faith to the meaning of human existence. The priest, as a leader in the Catholic community, cannot stand aloof from the quandaries that contemporary life presents. And one of the major questions that all priests must face (for themselves as well as for others) is: 'What is the meaning of my life and where is it leading me?' In the Catholic tradition such questions cannot be answered simply from biblical revelation or by some easy fundamentalist panacea. The Catholic church has a long history of reflection upon these issues and the priest needs to know that history, as well as knowing how to articulate it in today's culture.

This is why the dropping of philosophy from seminary studies was a serious mistake. Even to tackle basic contemporary theological texts, such as Karl Rahner's *Foundations of Christian Faith*, one needs some philosophical training. Priests must be able see the bigger picture. True, people are not asking priests for abstract philosophical solutions to the major issues of life, but they are asking profound existential questions about meaning and purpose, questions that are, fundamentally, philosophical. Because many priests and ministers lack the ability to think systematically, they try to answer profound human questions with superficial psychology. In many ways today psychology has replaced philosophy and, in the process, the symptoms have become confused with the cause. In my pastoral experience, people do not go to priests for psychological counselling; they go to talk about much deeper issues that are, at heart, philosophical, theological and spiritual. I am not suggesting that priests do not need to know some psychology, especially the classics, such as Jung and Freud. But as Jung so wisely commented, he had never known a psychological problem that was not, at heart, spiritual! And it is the specific job of the priest to address the spiritual.

My insistence on the importance of philosophy arises from my deep conviction that Catholicism will abandon part of itself if it loses its traditional ability to hold faith and reason together in a creative tension. It is the ability to reconcile the two that constitutes the narrow line between humane religion and fanaticism.

The second important issue for the future of seminaries is the need to maintain a mix of priest and lay students. Seminaries must not revert to becoming clerical enclaves — 'priestly, solemn, celibate places' as Thomas Keneally calls them in *Three Cheers for the Paraclete* — in spite of pressure from Rome to push lay students out. Even from a purely pragmatic point of view the economics of running such large institutions requires a student body of sufficient size to make the presence of a considerable group of laity vital.

I want to emphasise the importance of the tertiary theological education of lay people. As they take on new roles and ministries in the church, they need professional training. This can pose a problem for conservative church leadership because the more lay people gain theological knowledge and sophistication the more they become frustrated with old-style clergy. As Neil Ormerod points out:

The age of the 'simple faithful' so beloved of conservative clergy is beginning to end as laity become at least as theologically sophisticated as their ministers. The existence of such people can be both a threat to clerical authority and a source of constant frustration for those so educated as they seek ways of making use of their expertise. [64]

Many of the most talented and best trained of these lay people are women. Lay students far outnumber priest students in most of the theological consortia, so their place seems assured for the future.

Finally, as the seminaries train priests for a church that will increasingly emphasise the plurality of ministries, an appropriate spirituality to underpin this approach will have to be developed.

---

64. Ormerod, art. cit., p. 9.

Priests today are faced with the difficult asceticism of handing over their power and much of their ministry to others in the church. This calls for a great deal of personal integration, detachment and willingness to support their replacements in ministry. This level of generous maturity does not just happen; it has to be developed. The priest's role of pastoral leadership also requires that priests support and encourage others in exercising their ministry. This too requires the maturity of not only allowing others space and freedom to exercise their gifts, but also the ability of drawing all these ministries together and giving them a sense of direction. It cannot be expected that candidates will simply 'have' this maturity. There needs to be reasonable expectation that they will be able to develop it and commitment of resources to enable them to do so.

We will now examine a closely connected new development in the Australian church — the two new Catholic universities.

## Catholic universities

Like it or not, Australian Catholics are going to have two Catholic universities at the beginning of the academic year 1992. From the start I want to say unequivocally that I support the idea of a Catholic university in Australia. But this does not mean that I do not have reservations about the two particular specimens that are at present struggling into life.

The idea of a Catholic university has been around for a long time. In the late 19th century Cardinal Moran of Sydney saw the seminary he established at Manly as a 'training place for the Catholic University of United Australia'. [65] Ronald Fogarty notes that 'In the years following World War II an attempt was made in Sydney to establish a Catholic university; but owing to formidable opposition on the part of the non-Catholic churches and the consequent failure of parliament to pass the enabling legislation, the attempt was frustrated'. [66] The site had actually been bought at Mona Vale and the Congregation of the Holy Cross (from Notre Dame University in the United States) had been asked to set up the university.

Fogarty also notes the strong opposition of Archbishop Mannix of Melbourne to the idea of a Sydney Catholic university. While there was probably an element of Melbourne-Sydney rivalry in this opposition, Mannix pointed out — correctly in my view — that

---

65. Quoted on O'Farrell, op. cit., p. 267.
66. Fogarty, op. cit., Vol II, pp. 449–450.

Catholics should move out into the wider society of Australia and influence the culture by a full participation in national life. In the terms of this book, Mannix opposed a sub-cultural Catholicism which he believed a Catholic university could easily involve. He also felt that a Catholic university would mean that Catholics would abandon the existing state universities to the secularists. His criticisms of a Catholic university are well worth keeping mind in the contemporary debate.

Today, while circumstances are very different, taking the step of establishing two Catholic universities is still a radical move. It signals a major change in Catholic attitudes and practice regarding tertiary education. Catholic lay students are very much part of the ordinary university scene, although on many campuses they are seriously neglected in terms of pastoral care. Also, as I have just pointed out, Catholic faculties of theology, which so far have been the church's major tertiary institutions at university level, have been strongly committed to an ecumenical context for theology. Suddenly the ballgame has changed radically, with the unexpected establishment of *two* Catholic universities. The east coast Australian Catholic University (ACU) began operations at the beginning of 1991. Notre Dame Australia (NDA) in Perth begins with the academic year 1992.

These two institutions are very different from each other in the style of their operations, their origins and their evolution. Notre Dame Australia is a privately funded university, designed to operate on one campus in Fremantle, Western Australia; it is a totally new venture. In contrast, ACU is a publically funded university that will operate as part of the new unified national system. It will be based on eight different, already established campuses in five cities (Brisbane, Sydney, Canberra, Melbourne and Ballarat) and will draw on the existing staff and resources of the Catholic colleges of advanced education which the university has replaced.

## Notre Dame Australia

The history of the evolution of the two universities is both interesting and significant. Notre Dame Australia was first proposed by Perth Catholic businessman, Denis Horgan, chairman of the Barrack House Group, in the late 1970s. In 1986, the then Director of Catholic Education in Western Australia, Dr Peter Tannock, suggested that Horgan support the establishment of a Catholic teacher's college and the proposal received the blessing of the then premier, Brian Burke. Out of discussions on the teacher's college came the proposal for a

Catholic university. The university itself says that the proposal came about 'as a result of an initiative of a group of West Australian Catholic lay people'. Early on, the late Archbishop William Foley of Perth seems to have kept his distance from the group developing the idea; his position was not clear until late in the development of NDA. A proposal arose which involved a filial relationship with the United States University of Notre Dame at South Bend, Indiana, hence the title 'Notre Dame Australia'. After examination of the proposal by the Americans, the support of Notre Dame in South Bend was eventully obtained. This is how NDA explains the relationship with Notre Dame in the US:

The University of Notre Dame will give guidance during NDA's foundation years and it is intended that there will be a continuing association between the two universities. The universities are autonomous, and the relationship recognises their clear differences of culture and tradition. However, it places high value on professional and personal links between the two institutions, and their shared mission. Important factors in the association will be staff and student interchanges and advice on academic programmes and standards. [67]

The NDA *Overview* claims that the South Bend Notre Dame 'is ranked among the top American universities'. That is not strictly true. It is certainly recognised as one of the leading *Catholic* universities in the United States. But to see it ranking with Ivy League universities like Harvard or Princeton, or leading state universities, such as the University of California, is quite incorrect. It would be more accurate to say that it is one of the better second level institutions in the United States.

In August 1988 agreement in principle to set up NDA was announced jointly by Archbishop Foley and the then Deputy Premier, David Parker. Land and buildings had already been obtained in Fremantle in preparation for the first intake of 400 students who according to NDA 'will be drawn from all Australian states and from nations throughout the Asia Pacific region'. The exact role of the Burke Labor government in the setting up of NDA is unclear — as is so much about that particular state government! What is now clear is that the Labor government made a considerable land grant to this private university.

In Perth there has been ambivalence about NDA. Professor

---

67. *The University of Notre Dame Australia: An Overview*, 1990, p. 4.

Peter Boyce, Vice-Chancellor of Murdoch University, raised the issue of church/state relationship by arguing that the government had granted special privileges to the Catholic church; he also said that an explicitly Catholic university would be divisive. Boyce was accused of sectarianism and it was pointed out that he was probably upset because NDA might take students from Murdoch. Criticisms have also come from within the church itself. Some Catholics fear that money for NDA will be drained off from the needs of Catholic primary and secondary education to support what is a purely private venture. Sister Veronica Brady, a Professor at the University of Western Australia, made the telling point that NDA would form a kind of Catholic tertiary sub-culture, divorced from the place where Catholics should be: in the mainstream of state funded tertiary education. Attention has also been drawn to the NDA *Overview*'s reference to students from 'the Asia-Pacific region'. There is a fear that NDA will not be able to recruit many Australians because of high fees and will then be forced take in rich Asian students to fill up its places. In that way it will become something of a foreign enclave. Certainly the fee structure at NDA will have to be very high if there is to be no government money invested in the venture. This is how NDA explains the sources of its financial support:

NDA will depend for much of its ongoing financial support on tuition fees, private fundraising, endowments and Church and community contributions. From the State Government, the university will seek some establishment support, including access to State owned sites, and possibly loan guarantees, low interest loans and taxation concessions such as apply to other Australian universities . . . Interests associated with the university have also purchased further buildings in Fremantle with the intention of providing commercial support for the university. [68]

In Perth, most people are not opposed to a Catholic university as such, but are concerned that this particular venture is very much the product of an in-group who pushed the idea through. As I mentioned, even the late Archbishop Foley seems to have been quite reserved toward the idea until very late in its development.

One of the benefits of the university, in my view, is that it will reintroduce philosophy and theology into general undergraduate education. NDA says clearly that there 'will be a commitment to studies in philosophy and theology'. It has followed the US educational

---

68. *The University of Notre Dame Australia: An Overview*, p. 4.

pattern by demanding that undergraduates take a broad education beyond their major focus of study, included in which are four compulsory units in philosophy and theology. This US pattern is a major improvement on the Australian university system, whereby students are allowed far too much specialisation too early. In this matter the influence of the better American universities can only improve the general education of Australian students.

## Australian Catholic University

Australian Catholic University has a quite different genesis. It is the result of the amalgation of four Catholic colleges of advanced education. These colleges had evolved from the Catholic teachers colleges that were set up initially to prepare members of religious orders for the teaching profession. These training institutes, or 'scholasticates' as they were sometimes called, in turn evolved into colleges to train both religious *and* lay teachers. For example, the three Sydney campuses of ACU began as educational training centres for the Brown Josephite Sisters (in North Sydney), the Christian Brothers (in Strathfield) and the De la Salle Brothers (at Castle Hill). Signadou College in Canberra began as a training centre for the Dominican Sisters and McCauley College began as a training centre for the Brisbane Mercy Sisters. 'By 1980 the student body and teaching staff [of these colleges] were predominantly lay'. [69] Nursing was introduced as a subject into the Catholic colleges of advanced education in the mid 1980s, coming to them from the Catholic hospitals, many of which have been training nurses for most of this century.

The immediate genesis of ACU is to be found in the Dawkins reforms of higher education in Australia. It quickly became clear that the small Catholic colleges of advanced education would have no future in the Dawkins scenario unless they amalgamated; to continue to receive government funding an institution was required to have at least 5000 students. Only by getting together on a national level could the Catholic CAEs survive. As one of the proponents of ACU, Brother Dan Stewart says clearly:

With the ongoing reforms in tertiary education and the accompanying demise of the two-tier system of universities and colleges, the Catholic and

---

69. Stewart, Dan: 'A Case for Australian Catholic University', *Australasian Catholic Record*, 48(1991), p. 8.

other private colleges faced the choice of remaining separate, independent colleges or of joining the Unified National System. To choose the former would have been institutional euthanasia. [70]

Realising this, senior Catholic advanced college staff made the radical decision to pool their resources and students on a national level and move to university status.

Two things are clear from this. Firstly, ACU was primarily Dawkins-driven. In other words, the idea of ACU did not come from some ideal of Catholic tertiary education; it emerged from a pragmatic situation. It was the federal government, not the Catholic church, that made the fundamental decision and the Catholic colleges responded to that state initiative. This is a vivid example of the way in which the piper calls the tune; government funding of church institutions means that governments will determine the fundamental principles and the timing by which things happen.

Secondly, the decision to establish ACU (like the decision to establish NDA) was made by a small in-group of tertiary administrators and their advisers, with the eventual blessing and support of the key bishops. A major church decision was thus made without any discussion by or consultation with the wider church. Unless they were 'in the know', even the best-informed Catholics did not comprehend clearly what was going on until a few months before it actually happened. There were no discussion papers, no public consultation and no possibility for debate. As Father John Hill, a former Principal of the Catholic Institute of Sydney, said recently:

The Australian Catholic University, if its name is to be taken seriously, concerns the Catholic church in Australia . . . and not just the highly qualified and well meaning people who know the real direction of the project and resent criticism from 'outsiders'. There are no outsiders in this. The project concerns all Australian Catholics, who will be asked to support it, financially and otherwise. If they are ill-informed that is hardly their fault. They have questions, and they are entitled to some answers. [71]

In this same context, I commented recently on the foundation of ACU in *National Outlook*: 'It is simply another example of the unconscious assumption that a small group of bishops and their

---

70. Stewart, art. cit., p. 8.
71. Hill, John, 'University? Catholic?', *Australasian Catholic Record*, 48(1991), p. 12.

advisers "own" the church, and the rest of us are simply "disciples". [72]

The *Mission Statement* of ACU contains all of the expected university ideals about fostering research and scholarship in the widest sense 'in accordance with Christian principles and traditions'. [73] In his speech launching ACU, the Pro-Chancellor, Bishop George Pell, spoke of the university as an expression of 'Christian humanism', where, he said, the 'consequences of liberal education in a free and democratic state' would be able to be expressed. A statement like this is usually seen as indicating a fairly broad approach to scholarship in the tradition of Catholics like the 16th century scholar Erasmus and, in our day, the great Jesuit scholars John Courtney Murray and Karl Rahner. In the very next breath, however, Pell was talking about enticing students 'to recognise and embrace the romance of orthodoxy'. [74] This might be perfectly acceptable, except, as we have already seen, Bishop Pell has a rather narrow definition of orthodoxy. So much for 'Christian humanism'!

The *Mission Statement* goes on to emphasise that as 'a Catholic institution [ACU] establishes its corporate identity through the profession and study of the Catholic faith, tradition and doctrine'. It may be an oversight, but it is interesting that the word 'theology' is not used in this context. If theology is to be taught seriously at ACU, some very hard work will have to be done on staff recruitment. It is true that individual lecturers have been active in research and serious theology. On the positive side, the recruitment of a faculty of theology may give ACU a chance to move in a new direction from what has already been established. It may also provide places for the qualified laity now emerging from local faculties of theology.

Whatever way it goes, ACU needs to keep a clear distinction in mind between academic theology and religious education. As Karl Rahner pointed out years ago in the context of seminary education, proclamation and teaching are different realities from the essentially speculative nature of theology. Different methodologies are required. ACU needs to make sure that the distinction is clearly maintained. The former colleges that now make up ACU have — correctly in my view — specialised in religious education. But they have yet to

---

72. Collins, Paul, 'Do Church Leaders Talk to the Real World or to Themselves?', *National Outlook*, 13/2(1991), p. 21.
73. Stewart (art. cit., p. 10) quotes the mission statement in full.
74. Pell, George, 'Australian Catholic University. Inaugural Address', *Australasian Catholic Record*, 48(1991), p. 5.

establish any tradition of serious theology. Also, ACU has no tradition of philosophy, which, I have already argued, is essential for serious theology.

It is significant that all the major Australian Catholic faculties of theology have so far remained outside the development of ACU. As well as their commitment to theology in an ecumenical context, part of the reason for not participating is that the strongest theological tradition in Australia is to be found in the clerical religious orders and in the seminaries. With a few individual exceptions, there has been no real tradition of academic theology in the teaching orders. I say this not to criticise or blame them; it is simply to state the fact. And it is the teaching orders that constitute many of the key people involved in the formation of ACU. An important issue for the future will be the need for ACU to tap somehow into the already well-established Australian Catholic theological tradition. Without serious theology, ACU could hardly be called a 'Catholic' university.

One of the most frequent criticisms of Catholic universities is that they lack freedom of enquiry because they are dominated by the dogmatic concerns of the church and the theological preoccupations of the Vatican. The recent case of the sacking of Father Charles Curran at the Catholic University of America is often cited as a warning — although this university is not strictly a parallel to ACU. The Catholic University of America is a pontifical university, that is, its theology degrees are granted under a charter from the Vatican.

In fact, there are several distinct types of Catholic universities. There are the great historical Catholic universities of Europe (like the Catholic University of Leuven, Belgium) whose status rests upon their long historical tradition; there are a number of pontifical faculties (like Catholic University of America); and there are many independent Catholic universities usually founded by specific religious orders (like most of the North American Catholic universities) with their degree-giving status recognised by an outside, non-church authority. And there are now the new developments like ACU and NDA. The freedom of these two universities from interference by the bishops needs to be guaranteed, so that they can build a reputation for independent serious theological and other forms of research. ACU is especially vulnerable as it is not yet clear what exactly its relationship to the bishops is and how its senate will be composed. All types of questions about the teaching and composition of the theological faculty pose themselves: Who will hire staff and under what circumstances will they be fired? What rights will they have to pursue independent research? What limits will be placed on episcopal supervision of the faculty? The secretive way in which ACU was suddenly established does not bode well for responses to these questions, but they must be answered if ACU is to build any reputation for objectivity and academic freedom.

One of the most trenchant criticisms of ACU has come from Father John Hill of the diocese of Broken Bay. I have already referred to his comments about the secrecy surrounding the establishment of the university. In his article in the *Australasian Catholic Record* he identifies key problems associated with the foundaticn of ACU. He emphasises that he is not opposed to a Catholic university as such; his criticisms are directed specifically at ACU. He argues that ACU will be neither 'Catholic' nor a 'university'. He says that there are four elements in any genuine university: it is a community of scholars, who work interactively from their different disciplines toward a sense of the unity of all knowledge. They feel that truth is worth pursuing for its own sake and they believe that the university must be counter-cultural — a genuine critic of the society of which it is part. This last point is the focus of Hill's concern. He concedes that because it comes from within the context of the Catholic tradition, ACU may have a chance of developing counter-cultural attitudes, but he points out that:

shortage of money, declining prestige, and growing dependence on government have made it difficult for universities to be counter-cultural . . . [However], if [ACU] exercises a role of critic of secular society from an explicitly Catholic standpoint, there will be some point in having it, but it will need to watch where government policy is leading it. Catholic hospitals learnt that lesson too late. The Catholic University is already at the bottom of the garden path. [75]

As I have said, I consider dependence on government money to be a key problem for all Catholic educational enterprises, but this is a special problem in view of the government initiated policy that is at the genesis of ACU.

The current climate in tertiary education could not be worse for the foundation for a new university, dominated as it is by the worst form of philistinism, initiated largely by the federal Labor government, the federal educational bureaucracy and their sundry advisers (mainly from universities), all presided over by the minister,

---

75. Hill, art. cit., pp. 14–15.

John Dawkins. Prime Minister Hawke's call for the development of the 'clever country' says it all. The very word 'clever' reveals the crude pragmatism at the heart of this philistine policy. Australia is not to be wise or intelligent but 'clever' — the word suggests an Australia that is pragmatic, street-wise, on the make, looking always for the quick buck and the easy money. The Hawke Labor government has a track record of being profoundly anti-cultural. Its record is even worse than that of successive Liberal governments and they were bad enough! Hawke Labor has been more destructive of the institutions that support and help make up the fabric of Australian culture than probably any other federal government since federation.

As John Hill says: 'The ideology behind the emerging Australian universities is materialistic, pragmatist and utilitarian'. There is no place for moral, intellectual, educational or cultural values. The dominant pragmatist-materialist ideology of today is more deeply secular and antithetical to Christianity generally and Catholicism specifically than the liberal secularism of the 19th century, when Australian Catholics turned away from an education system based on values incompatible with faith. Like the Catholic school system of the past, a genuinely Catholic university of today will have to stand against the prevailing ideology. ACU certainly faces a challenge in this. As Hill says: 'It will be a bold university that can explicitly resist the Australian orthodoxy that an effective pursuit of values is less important than an efficient preparation of future money earners, particularly if it pursues values that are opposed to materialism, pragmatism, utilitarianism and greed'. The problem, as Hill sees it, is that ACU will just not be brave enough to take the stand against the Dawkins educational monster and federal government philistinism.

Hill's final criticism is in the area of community. How can you have a community of scholars if the staff and students are situated on eight different campuses spread over five cities that are geographically and socially diverse. The geographical spread does mean that ACU has at least a partial chance of being 'national' although what people in South Australia, Western Australia and Tasmania think of this is another matter? The problems of the geographical dispersion of the campuses are obvious and it will require great commitment to overcome them.

In my view, the most important criticism of ACU and, for that matter of NDA, is that either of these institutions could easily create a Catholic intellectual sub-culture. This is especially dangerous at a time when our mainstream secular universities are under such a serious threat from philistinism. It would be a terrible loss if Catholicism retreated from the mainstream of intellectual life into a safe and theologically sanitised sub-culture. Of course, this argument can be turned on its head: both these universities could become bases from which Catholic academic life could reach out to the wider community. It will all depend on policies adopted now. If an early pattern of self-absorption and sub-culturalism is set, it will take years to break, and those Catholics who believe that the church has nothing to fear from the modern world will choose to study elsewhere. The danger is that a narrow and fearful mediocrity, or, even worse, a reactionary rump, will gain control. If ACU and NDA are to be accepted as part of Australian culture and participants in the Catholic mainstream, they must stand firmly in the tradition of the contemporary post Vatican II Church. Since there is no ecclesiastical 'left' in Australia, the greatest danger that ACU and NDA face will come from the reactionary Catholic right.

I do not want to leave the two Catholic universities without saying that I hope that they succeed and that my criticisms of them do not imply that I think that the idea is wrong. The North American Catholic tertiary education sector has made a very significant contribution to the church, to an educated laity and to American cultural and academic life. But the structure of North American Catholic tertiary education is very different from Australia's, especially the situation of ACU. Admittedly, NDA has more of the American spirit, where all church institutions are privately funded; but this requires real sacrifice from the Catholic community for their maintenance. The two universities will only contribute something significant to Australia if the church is prepared to commit talent and resources to them. That will require an unswerving commitment to intellectual freedom, liberal education and serious, untrammelled research. It is precisely silly statements like Bishop Pell's 'romance of orthodoxy' that will frighten off serious scholars. The genuinely orthodox Catholic tradition is a good deal more broadly humanistic than is conceived of by some of the bishops. If the Catholic universities are truly catholic in the breadth of their views and the depth of their scholarship, then they will have much to contribute to Australian culture and to the church. If they produce nothing more than a safe and narrow sub-culture, then they will be a waste of government money and the Catholic community's effort.

## Campus ministry

Finally, I want to turn to the ministry of the church on the existing secular university campuses. Until the late 1960s and early 1970s, the universities were cared for by able priest chaplains, a number of whom deeply influenced several generations of Catholics, some of whom are now leading lay Catholic thinkers. But post-Vatican II instability saw a number of the well-known priest chaplains leave the active ministry, and for a while many bishops and conservative Catholics saw a university priest as one 'on the way out'. No doubt this was exacerbated by the student revolt of the late 1960s and early 1970s and the strongly non-religious stance of many intellectuals. It is my impression that in this period the church lost its nerve on the campus.

It does not seem to have regained it in the 1980s. Catholic chaplaincies have a low profile on most university campuses. Part of this is caused by the attitude of the universities themselves who see campus ministry as a very low priority. They seem to think that the university counselling service has taken over the job of the chaplain. But the church itself is also to blame. Few dioceses are willing to put up enough money to support a campus ministry team of laity, religious and clergy. The North Americans have shown that the team approach is by far the best way into the modern university; it provides support that the isolated chaplain simply does not have.

There is little liaison and co-operation about campus ministry between religious orders, who could and sometimes do provide personnel, and the responsible dioceses. This has left the field wide open to all types of charismatic and fundamentalist groups, who have literally invaded the universities. The lack of a strong intellectual Catholic presence working to integrate faith, reason and knowledge allows impressionable students to be caught in the simplistic verities of the fundamentalists.

Campus ministry and university chaplaincy is at present in a vacuum that needs to be filled with commitment, intelligence and appropriate personnel, whether or not the Catholic universities succeed. The vast majority of Catholic students will continue to be on ordinary Australian university campuses for a long time to come. The church cannot simply neglect them in a key period of their intellectual and personal formation.

The Catholic universities are one form of new ministry. I want to look now at some of the other new forms of ministry that are emerging. Our discussions of various aspects of pastoral work have already indicated some ways in which ministry might develop in the future. However, there are a number of new ministries, or new ways of doing things, that are emerging to which we should pay attention.

## *New ministries in the Australian church*

Just before the demise of *The Catholic Worker* in 1989, another lay and non-official tabloid appeared called *Adelaide Voices*. *The Catholic Worker* had been produced by an ad hoc group of lay people in Melbourne. *Adelaide Voices* is written and edited by a much more highly organised group — the Romero Community, previously called the Sturt Street Community. This community has a core group of about nine people who live and work together. Associated with them and participating in their ministry, is another group of people, some of whom are single, some married with families. They are all committed to the practical ministry of a day centre and soup kitchen for homeless men; they also support El Salvadorean refugees and are committed to a regular celebration of the eucharist. Thus *Adelaide Voices* has a chance of survival that is lacking where a group is simply ad hoc.

The Romero Community is an example of a growing phenomenon in Australian Catholicism (and Australian Christianity generally): the growth of the local community/home church/small group movement. This arises clearly from the alienation that characterises modern western society. Sydney Dominican Sister, Mary Britt, has shown that what is happening in Australia in the small community area is simply part of a wider movement through western Christianity. [76] Many people are attracted to a community model of Christian life, but feel that the conventional commitment of religious life as it is presently constituted is not for them, at least not until the structures begin to adapt, and more flexibility is allowed. Many young people are unwilling to make a life-long commitment, but do want to give a period of their life to a committed ministry, with or without celibacy.

I know one highly intelligent young woman who, in the middle of an arts-law degree, committed a couple of years to the Jesuit

---

76. Britt, Mary, E., *In Search of New Wine-Skins: An Exploration of Models of Christian community*, Collins Dove, Melbourne, 1988.

Volunteers and worked with the homeless. When she returned to her legal studies, her social conscience was much more highly developed and the experience will profoundly influence her career development. It is this youthful idealism that is being tapped very successfully by evangelical Protestants such as Youth with a Mission. Many young people today with a willingness to commit themselves to ministry are the ones who, thirty years ago, would have entered religious orders. But not today; now they are more likely to make a commitment to a specific group or to a ministry for a fixed period of time. The most prophetic among these people are likely to step right into the firing line by joining a group like the US Catholic Worker community. An example of this level of commitment are the two young Australians — one Catholic, one Protestant — who were charged recently in the United States for a protest action on an air force base in northern New York state. Such people run great risks for, as many Christian pacifists like brothers Philip and Daniel Berrigan have found, the US military invoke the most draconian laws to protect their fighters and bombers from the 'danger' of prophets armed with nothing more than their bare hands and their own blood. No doubt, the military are also embarrassed by their own lax security, which allows two young Australians to sit on the runway of a secure air force base for most of the night, digging a tiny symbolic hole in the tarmac!

Britt points out that, just as people are dissatisfied with the affluent, self-indulgent democracy of the west, there is also considerable disenchantment with the institutional churches. [77] This disenchantment has arisen because the churches generally, and the Catholic church specifically, shy away from a prophetic, creative role in society. As established institutions they usually decline to rock the boat. Historically, the churches have backed away from a prophetic role in every society in which Christianity has been the traditionally predominant religion. Modern Australian secularism, unable to eliminate the religious urge in society entirely, attempts to confine the churches to an exclusively sacral role, which it then relegates to a secondary and ancillary status. Often the churches collude in this, for it is a comfortable and secure role: the church as chaplain to secular religion. The domesticated clergyman is rarely a prophet!

Mary Britt distinguishes four challenges that base communities try to confront: (1) They try to see reality from below, from the way

---

77. Britt, op. cit., pp. 7–8.

the lowest social strata see it — hence their commitment to the poor and the marginalised. (2) They try to read the bible from the perspective of the poor. (3) They try to find new ways of going about Christian mission; new ways of proclaiming the gospel and working for peace and justice. (4) They try to discover in their communities a new way of living in the church and a new model of the church. [78] The growth of these communities has been deeply influenced by liberation theology and the development of the ecclesial base communities in Latin America. However, the most sensible communities in Australia have realised that the reproduction of the Latin American model is impossible here, where the social and economic situation is so different. Liberation theology is not so much a theology, in the sense of having specific dogmatic content, but a methodology, a way of looking at society critically in the light of Christian and biblical revelation. Liberation theology is obviously deeply rooted in the methodology of Joseph Cardijn's YCW: *see* the reality, especially the reality of injustice and exploitation, in society around you; *judge* it through a process of discernment with fellow Christians in the light of the bible and the church's tradition and teaching; then *act* in the real world to change that evil reality, no matter what it is or whom it involves. Christians ought to be able to do this in Australia and stop worrying about doing liberation theology in the Latin American style.

There is a remarkable number of these communities and small groups in Australia. Some are formal, some informal; many are not even known to each other. A number, like the Grail, have been at work within the local Catholic church for a long time. The Grail came to Australia in 1936 and has provided a model to the church of dynamic, independent women, free of clerical control and the constraints of conventional religious life, working in ministry. [79] A core of trained and committed women forms the nucleus of the Grail, with other members establishing varying degrees of involvement with the movement. It is active in ministry in Sydney, Melbourne and Townsville.

One of the most successful of the new communities in Australia is L'Arche, founded in 1964 by Jean Vanier, an academic and member of the French Canadian élite; his father had been a Governor General

---

78. Britt, op. cit., p. 7.
79. For an account of the Grail in Australia see Sally Kennedy, *Faith and Feminism*, Studies in the Christian Movement, St Patrick's College, Manly, 1985.

of Canada. L'Arche is a community centred on care for the disabled. 'Fundamental to Vanier's approach to community is acknowledgement of the fact that we are all handicapped, wounded, weak, sinful but some handicaps are more obvious than others'. [80] A number of other communities are the offspring of the charismatic movement, often referred to as covenant communities — committed Christians who pledge themselves to follow the way of life of the community. They are sometimes ecumenical in membership. A number of commentators have pointed to the somewhat hierarchical and patriarchal character of the groups that have emerged from the charismatic movement, but generally speaking in Australia the Catholic communities have remained well within the bounds of the mainstream church.

A few of these new Australian communities are rather inward-looking and self-engrossed. I have referred to them in passing when discussing new developments in religious life in Australia. I think that there are probably several ways of discerning whether a specific group or community is fulfilling a genuine need. Traditionally in Christianity, community has always been directed ultimately toward the service of others; it does not exist for its own sake, or only for the sake of the members. In other words, it is oriented to ministry. Saint Paul says clearly that belief will always bear fruit in action. Nowadays, with the strong social justice emphasis of the Catholic church, this ministry should have a social justice perspective. An exclusive focus on personal spirituality, or, more likely, self-fulfillment, seems to me to run the risk of divorcing the contemplative life from contemporary social reality.

In the United States some Jesuit retreat houses are deliberately established right in the middle of the worst slums of the inner city. In Detroit, for instance, the Jesuits conduct retreats in an area of urban decay that is shocking even for the US. Retreats in Detroit have no problem focusing on social justice where unemployment, drug addiction and a very high rate of homicide are the product of the despair of black Americans. In this context, social justice and contemplation are profoundly integrated.

The search for community in a culture is, to me, a sign of the decline of that culture. When a culture is in good shape, people do not have to search for community, it exists naturally. The danger today is that what people are offered is an artificial, self-indulgent,

---

80. Britt, op. cit., p. 45.

merely therapeutic group. Genuine Christian community in our culture can redirect the narcissism inherent in much of the contemporary desire for social communion by harnessing the energy involved and directing it outwards towards other people. True communion is only found in shared self-giving.

I should not leave this topic without saying something about another movement closely related to the community movement: the growing network of household or home churches. In both the Protestant and Catholic churches the origins of these groups lies in dissatisfaction with large parishes and in a desire to build a smaller, interactive and participative fellowship. One of the major theorists of the movement is an Australian, Dr Robert Banks, one of the moving spirits of the Protestant home church movement in Canberra. [81] Banks sees the home church as a new expression of the gatherings of the New Testament and the first centuries of Christianity. At the heart of Bank's understanding is the concept of fellowship, by means of which people are able to transcend the anonymity of large parishes through participative interaction and a sense of mutual responsibility.

The house church movement is also to be found among Australian Catholics. One group I have encountered through their newsletter is called *Oikia*. It has a core group of families drawn from various parts of Victoria. A quote from US lay theologian, Michael Cowan, in *Oikia Newsletter*, makes clear the motivation behind the house church in Catholicism:

Many adult members of the community of faith . . . have begun to feel the need for and the power of gathering with other adults in groups of a size which permits personal sharing of the struggle to order our lives in this highly individualistic and materialistic culture in fidelity to the Christian tradition. The typical Sunday liturgical assembly, while crucial to the integrity and vitality of any parish, is simply too large to permit the level of personal conversation with scripture, the later tradition and other persons that is required for a mature and integrated adult life of faith. [82]

Within Catholicism, groups like this have not been divisive in any way. Most members of house churches maintain their contact with the mainstream church on a number of levels. Most of them already have a network of relationships within the Catholic community, their

---

81. Banks, Robert and Julia, *The Home Church*, Albatross, Sydney, 1987.
82. Quoted in *Oikia Newsletter*, Christmas 1989.

children go to Catholic schools and they maintain an interest in all that is going on in the church. It is from groups such as these that a renewed vitality for the local church can emerge.

A major problem facing the various lay groups in the church is that they tend to be very localised, without a diocesan interconnection that is not controlled by the hierarchy. This is true also across the dioceses. That is why the Australian Lay Network was a very far-sighted venture when, in its own words, it 'began to emerge in 1984. It came from the strong desire of a number of lay people to more properly participate in the life of the church and to share that opportunity through mutual support with other lay people'. Significantly, the Australian Lay Network was set up to be 'independent of existing church structures and bodies, so that it may better enable the emergence of an authentic lay voice, prophetic in nature and concerned with the needs of the entire church'. [83] There is a real need for such an independent lay network, so that another voice can emerge, and the energy and commitment of laity can be channelled in ways that are not dependent on and controlled by the bishops. The reference to the prophetic is significant; a lay voice that is channelled through the official church is unlikely to be 'prophetic in nature'. The Lay Network is now in abeyance, but the idea is a good one for the future and the people who started it are still in place.

## The media

There is one other area of ministerial initiative that I want to mention: the relationship of the church to the media. I have already mentioned this in an earlier chapter when talking about the Bishops Conference's failure to deal effectively with the media. Yet this area of human culture is increasingly important in our information-saturated age.

I will start with television and radio. Both these media are far too expensive at present for the churches, either singly or even on an ecumenical basis, to own and direct. Certainly, a church could (and some churches do) own individual radio stations, and with the advent of cable television, it may be possible to break into this hyper-expensive medium. But at the present moment entry into mainstream media is prohibitively expensive. The church will have to adapt itself to what is on offer from the ABC, SBS and commercial stations.

The ABC has a major commitment to Australian religion in

---

83. Quoted from the Network's newsletter *Grapevine*, December 1989.

television and especially in radio. Having worked with the ABC in both media for several years, I have had a chance to see the church's performance. The Catholic church, with notable exceptions in Melbourne and Adelaide and to a lesser extent in Sydney, does not take the media seriously. Radio and TV seem to be seen as the 'enemy', trying to get at the church or the bishops. It should be noted that this attitude arises from the media's own ignorance rather than any antagonism; most journalists know nothing about religion and the unsophisticated and defensive attitudes of churchpeople toward the media tends to reinforce the negative impression of irrelevance that most journalists have of the church. Many of them are surprised when they find that the church actually has a reasonable point of view and something sensible to say.

Where the church has set out, as in Adelaide, to relate professionally to the local media, attitudes begin to change. Perception is an important factor here; if the church is seen as closed and defensive, journalists will be out to see what 'dirt' they can discover. There is nothing more challenging to the media than defensive people who look like they have something to hide — even if they do not, and their defensiveness arises from nervousness! If the church takes the initiative and begin to use the right people to explain its teachings, moral attitudes and opinions, then a fruitful relationship with the media can be built. The right person is central to this effort. The church needs a group of men — *and* women — who can speak for it, both on a national and a local level. The situation at the present moment is frankly bad. The bishops have to accept that only a couple of them are really capable of talking to the media, and just because a priest is considered a 'safe' spokesman, it does not mean that he is good working with radio, television and newspapers.

Churchmen (the spokespersons are almost all men) often commit three serious sins in dealing with media: (1) They convey a sense of superciliousness because they do not concede that there may be another opinion besides their own or that of the church; they respect neither their interlocutors, nor the viewers or listeners. (2) They are too serious, forgetting that the media must have a strong element of entertainment and even the most serious subject should not be reduced to the deadly dull. A sense of humour, a ready smile and a willingness to laugh are essential for any media performer. (3) Media is a performance. In the trade, interviewees are called 'talent' and the talent that church representatives need, especially when they are 'talking heads', is the talent to perform. This requires an output of energy, a facility to think and speak on your feet and an ability to be able to explain simply and clearly the Catholic perspective. It is obvious that not everyone can fulfill these specifications; the church needs trained people who can be real 'talent'. In the area of radio and television the church has a long way to go before it begins to communicate effectively.

There *is* one area of media where the Australian church has offered a coherent and positive critique: film criticism. It is easy to underestimate the importance of films, especially since the widespread introduction of the home video. The success of the church in film criticism is due largely to the work of two men, Fathers Fred Chamberlain and Peter Malone. Father Chamberlain began film criticism in the early 1950s with two pamphlets *Films and You* and *You and the Movies*. By 1971 he had set up a National Catholic Film Office. Father Peter Malone began reviewing films in 1968 in the magazine, *The Annals*. In 1971 he published *The Film*, in 1973 and 1975 *Movie Discussions I* and *II* and *Films and Values* in 1978. [84] He has also written thematic works about nuclear war and Aborigines. In 1988 came *Movie Christs*, a book about Christ figures in film. [85] To help viewers, or parents and teachers, judge the moral and cultural worth of specific movies. Malone published *The Australian Video Guide* in 1990. [86] Malone's approach has not been limited to judgements about specific films; he has also reflected on the cultural and even theological role of movies in contemporary society.

We turn now briefly to the print medium. The recent dramatic closure of the Melbourne archdiocesan weekly, *The Advocate*, after over a century of publication, shows the vulnerability of the official Catholic media to the vagaries of ecclesiastical decision-making. The ostensible reason for the closure was the financial loss made by the paper and its decreasing circulation. It has been asserted by some critics that the real reason was archdiocesan distaste with editorial policy. At one period Melbourne Catholics had supported two diocesan newspapers, *The Advocate* and *The Tribune*, plus the journals *News Weekly* and *The Catholic Worker* and some smaller periodicals. Now all that is left of the official Catholic press in Australia, with any wide distribution, is Sydney's *Catholic Weekly*

---

84. These books were all published by Chevalier Press, Kensington, NSW.
85. Published by Parish Ministry Publications Marsfield, NSW.
86. Published by Collins Dove, Blackburn, Vic.

and Brisbane's *Catholic Leader*. There are small officially controlled publications in a number of other dioceses, but none with any wide circulation or breadth of coverage.

The official Catholic press, despite valiant efforts by many of the journalists involved, is generally dull and not particularly informative. It tells little about the most interesting issues, whether local news or overseas stories. You do get the 'official story' in these papers, but this is often not the whole story. There is a sanctimonious prissiness in the official Catholic press that conceals the real debates that are going on in the church. It is simply beyond me that the official church is so fearful of new ideas, and the position of church authorities will continue to be compromised until open debate within the structures of the church is allowed. It is heartbreaking to see how dull the official press is, when it could be so interesting, if the fascinating and passionate debates which engage Catholics were allowed a public airing. It would give the 'simple faithful' a chance to come alive!

A number of unofficial Catholic newspapers, magazines and periodicals have begun recently, or have been publishing for some time. They give expression to a broad range of views. Among weekly or monthly papers there is the National Civic Council's *News Weekly*, which has survived with a consistent editorial policy for many years. *Annals Australia* has had a much longer history and was a pioneer in developing a very helpful catechetical supplement from the 1960s onwards. The religious orders also get out a number of other smaller devotional and educational magazines. *National Outlook*, though ecumenical in its focus, has a substantial Catholic component. It has now been publishing for over 13 years, but it has only a limited circulation. As I have mentioned, *The Catholic Worker* had a short-lived revival. It has been more or less replaced by *Adelaide Voices*. A new arrival is the glossy magazine *AD 2000*, which is an outlet for the B.A Santamaria brand of Catholic conservatism.

In the area of serious religious periodicals, the long-lived *Australasian Catholic Record* has taken on a new lease of life under its present editorial board. *Compass Theology Review*, produced by the Missionaries of the Sacred Heart since 1965, has tried to bring theology to a wider readership. Recently the theological periodical *Pacifica* began publication; it could well become Australia's premier theological quarterly. Early in 1991, Jesuit Publications, with assistance from the Sisters of Mercy, the Loreto Sisters and the De la Salle Brothers, began the production of *Eureka Street*, a broadly cultural periodical with a Catholic leaning. It is too early to assess its impact.

Australian Catholics have always maintained a small but vigorous periodical press and this looks like continuing. Its impact is numerically limited, but its main function is to nourish the church's intellectual and ministerial group with ideas and debate — the type of debate that is so sedulously avoided in the official Catholic press. But there is something precious and rarified about much Catholic periodical writing. It is still a long way from engaging with issues that concern ordinary Australians. While our best writers have been exploring the concept of 'Australian theology', we have had a federal government systematically tearing apart the cultural life of the country, with attacks on the CSIRO, the ABC and the universities. I have read little about this in serious Catholic periodicals, nor much about that other central moral and social issue of the 1990s — the survival of the Australian environment. It seems to me that in the coming decade Australian Catholic discussion in intellectual and theological publications will have to come down to earth a little more, and engage much more directly with the moral and theological issues that most affect Australian society and culture.

❖ 4 ❖

# THE FUTURE OF THE CHURCH

In this final chapter I want to address the question: What will be the major issues facing Australian Catholicism in the next ten years as we proceed to a new millenium? I will look at these issues firstly from an intramural perspective — what are the important internal priorities facing the church? — and then from an extramural perspective, as the church looks outward at the major issues that face Australian society.

## *The church looks inward*

The first major intramural issue for the next few decades for the Australian church will be to continue the emphasis on adult formation. The focus of this formation will be on the rediscovery of the freedom of the Catholic conscience, within the context of a renewal of a genuine sense of tradition. Freedom of conscience and a sense of tradition will be the key elements in the spirituality of Catholics of the immediate future.

The second issue will be facing up to the difficult question of applying Catholic social justice teaching to Australian society. In my view, three key social justice areas already confront the Catholic community: the need for a critique of economic rationalism, the position of Aborigines and the issue of immigration.

## Freedom of conscience

One of the greatest successes of the contemporary Australian Catholic church, is adult formation. Slowly, the church has produced a cadre of adult laity who have living faith and a deep commitment to Catholicism. The church needs to continue this development, for it is as important and revolutionary a change as the establishment of the Catholic school system in the 1870s. At the core of this formation is the development of the centrality of freedom of conscience as the ultimate guiding norm and central spiritual focus of the adult Catholic Christian. Conscience is not merely a moral facility to judge right and wrong; it is the focal point of lay Catholic spirituality.

In the light of this, I would argue that Bishop George Pell made a quite extraordinary statement in his seminar on *Catholicism in Australia* at Latrobe University in 1988. He said: 'The doctrine of the primacy of conscience should be quietly ditched, at least in our schools, because too many Catholic youngsters have concluded that values are personal inventions, that we can paint our moral pictures any way we choose'. [1] Now it is understandable that the bishop, and other responsible adults, should be concerned if young people abuse their freedom. But the abuse of freedom of conscience by the immature young surely does not mean that the doctrine 'should be quietly ditched'. Perhaps a more mature response would be to work harder at developing a more responsible freedom of conscience. I am shocked that Pell, who told the distinguished gathering at the launching of Australian Catholic University that he delights 'in the romance of orthodoxy', should suggest the quiet 'ditching' of a teaching that he admits himself is a doctrine of the church! What religious education actually needs to do is focus more fully on developing in young Catholics the very element that will be the pivot of their adult spirituality, their conscience. However, the real issue for the wider church is not Bishop Pell's attempts to 'ditch' church teaching, but the fact that primacy of conscience is an important part of Catholic teaching, and that it will increasingly become a central element in the development of adult Catholic spirituality.

What do I mean by conscience? Before attempting an answer I must confess that I have always been suspicious of moralists generally

---

1. Pell, *Conversazione: Catholicism in Australia,* op. cit., p. 10.

and moral theology specifically. The reason, no doubt, is that I never did very well in the subject in the seminary, but I have never been able to rid myself of the suspicion that morality is not all that complicated and that any sensible Christian could work it out through prayer, reflection on the bible and the Catholic tradition, some sensible human advice and the courage to decide for oneself. Not that I want to trivialise the trauma and pain of moral decision making in important issues, but I do want to stress the necessity of discovering anew a genuinely Catholic freedom and autonomy. No one can foist onto others, or allow others to assume, responsibility for one's own personal, moral decisions. And to make an important moral decision for oneself does require courage: the courage to assume responsibility for our own lives. But, as we shall see, the term 'moral life' can be misleading, for we are really talking about the Christian life itself. Morality is not something independent of living a mature Christian spirituality.

Let us begin by examining a specific case to illustrate what I mean by conscience: the choice of Father Fernando Cardenal to take a personal conscientious decision that eventually led to his expulsion from the Jesuits. After the Sandinista revolution in Nicaragua, Cardenal was appointed minister for education and was the man responsible for developing the very successful literacy programme that taught most of the population to read in a country that, before the revolution, was largely illiterate. In 1984 he was ordered by the Jesuit Superior General, Father Hans-Peter Kolvenbach, under pressure from the Nicaraguan bishops, the Vatican, and ultimately Pope John Paul II, to resign his portfolio in the Sandinista government. Cardenal refused to do so and, after thirty-two years in the Jesuit order, he was expelled.

There is nothing new about the political machinations of popes, bishops, Jesuits and the Vatican. But the reasons for Cardenal's refusal to obey them are interesting, if not particularly new. In a 'Letter to my friends', he explains his reasons for refusing to obey:

I thought that I could nourish the hope that the church would see in my work a missionary type of apostolic service . . . for the poor. Because of this I thought I would be able to keep hoping that a conflict between a . . . command of the church and my conscience would never arise . . . I have dedicated a lot of time to discernment and spiritual direction . . . I have reflected on the whole of my situation with people of deep spiritual experience . . . Because of this, I can responsibly state that honestly, objectively and seriously I conscientiously object to the pressures of the ecclesiastical authorities. In all sincerity I consider that, before God, I would be committing a serious sin if, in the present circumstances, I were to abandon my priestly option for the poor . . . My conscience grasps, as if in a global intuition, that my commitment to the cause of the poor comes from God. [2]

In this Cardenal joins other Christians, like Saint Joan of Arc, who have made stands against ecclesiastical authority, in order to retain the integrity and primacy of their conscience, or what Saint Joan called her 'inner voice'.

What Cardenal has done for us in this passage is to state clearly the essence of conscience. Firstly, it is a total moral act that involves the whole Christian person. It is not merely an evaluative judgement that this or that action may be right or wrong. Conscience involves the whole person in making choices about fundamental commitments. Cardenal's conflict is between loyalty to the Society of Jesus and to the vows he made as a young man, and loyalty to his personal, radical commitment to the poor of Nicaragua, a commitment made as a mature man which led to his discovery of God in the poor. It is a question of discernment between conflicting loyalties and seeking to obey the loyalty which is most fundamental.

It is still foreign for Catholics to think of conscience in this way. Like many other key words, 'conscience' comes into contemporary theology loaded with historical connotations. Since the latter part of the 17th century, the word has been used primarily to refer to the faculty of moral evaluation. In this understanding, conscience focuses on specific actions and answers the question: is this act or course of action morally right or wrong? Thus, the contemporary church has inherited an idea of conscience that is linked closely to the determination and extent of sin and moral culpability. Linked to this is another element in the moral theological tradition. The 17th and 18th centuries were, in some ways, morally fear-filled times. Conscientious people asked 'How can you be certain that your action is morally right'? And the answer was: 'It will be right if it conforms to the law, divine or ecclesiastical'. As a result, the theological stress was placed on obedience rather than freedom.

The emphasis on obedience was strengthened by the influence of the moral theories of the great German philosopher, Immanuel Kant (1724–1804), whose opinions have deeply impregnated the whole European tradition. According to Kant morality did not come from

---

2. Quoted in the *National Catholic Reporter*, 21/11(1985), pp. 1 and 6.

God or the church, but from the intuition of universal moral norms. He played down the importance of individual conscience and emphasised the universal validity of the moral law. He formulated the principle: 'So act that the maxim of your will could always hold at the same time as the principle of a universal legislation' (*Critique of Practical Reason*, 5:35). For Kant conscience was merely the syllogistic application of this law. He argued that we experience an absolute obligation to obey the moral law — we find ourselves governed by the word 'ought' — *we ought to do our duty*. This expresses itself through a whole series of imperatives which are subsumed under what he calls the 'categorical imperative'. Fundamentally, this means that we ought to act in a way that our action can become a universal moral norm which can be a model for others. Rooted in the Kantian approach is a strong emphasis on moral perfectionism. Despite the rejection of Kantian philosophy, the seductions of moral perfectionism crept into Catholic attitudes, largely through preaching and catechetical instruction.

In the 19th and early 20th centuries morality came, in practice, to be identified with two issues: sexuality and obedience to moral and church law. In theory, there was a strong emphasis from the late 19th century onwards on social justice, especially following the pontificate of Pope Leo XIII, but at the pastoral level of everyday Catholic life, sex and obedience to law were the prime focus of morality. Moral autonomy had almost disappeared and sexual morality admitted no 'parvity of matter' — in other words, each and every sexual sin was equally and appallingly mortal! Freedom of conscience was still recognised in theory, but in practice it was swallowed up by the pervasive power of the church, exercised through the confessional. In the 1960s this 'old' view came more and more under fire in the church. Then came 1968 and the encyclical *Humanae vitae*. It clearly condemned all forms of artificial contraception. Given that many Catholics were already using the contraceptive pill, expecting that the pope would approve it, it was inevitable that the encyclical would cause a storm of protest and a major rupture in the Catholic experience.[3] The question was asked: How binding was the pope's teaching against the pill? What degree of obedience could it demand of Catholics? I actually heard a cardinal of the Roman Church say on Australian national television

---

3. For a history of the papal commission on birth control see Kaiser, Robert Blair, op. cit.

that the encyclical was 'almost infallible'! How something could be 'almost' absolutely true still puzzles me. Since there was no unanimity among moral experts in the church about *Humanae vitae*, Catholics of fertile age (and the priests advising them) had to make their own decision — and many, in doing so, attained moral autonomy in one frightening step.

There is a further and, in my view, deeper aspect of the crisis of the 'traditional' view of obedience and sexuality. I refer to the decline in the use of the confessional as a means of forgiveness and reconciliation in the church. In the previous few centuries, through a well-regimented clergy, the church exercised an enormous moral force in individual lives through the sacrament of confession. Those who know the secrets of the heart have tremendous persuasive power. But this is no longer true, for Catholics generally do not go to confession. The reason is not that people have lost their sense of sin; it is simply that telling their sins to a priest is no longer a credible way, for most Catholics, of receiving God's forgivness. In this context, it is important to recall that the sacramental practice of confession or reconciliation has gone through a very variegated evolution in the course of church history. However, the human need for sacramental reconciliation and absolution is still strong. This is why penitential services, especially those with general absolution, are so popular, both overseas and in Australia. This is not seen by people as an easy way out. It is the contemporary assertion of the profound and traditional Catholic consciousness that the real nature of human sin is much deeper and broader than a catalogue of specific actions. Sin has to do with the whole direction of a person's life, with the inner core of who they are as human individuals.

## 'With all your heart'

So where do Catholics and the church go from here and what is the importance of all this for the future? Firstly, there is a need to return to the roots of the Catholic moral tradition, especially to the bible. In the Hebrew scriptures, the 'heart' is the locus of moral decision. 'Heart' is also the word that is used to describe the whole religious commitment of the person: the psalmist asks God 'to create a clean heart in me' (Psalm 51:10) and the prophet Ezekiel asks God to 'take out . . . the heart of stone and . . . give . . . a heart of flesh' (Ezekiel 36:35). The heart is the radical centre of religious commitment. Hebrew theology does not focus on individual moral acts, but attempts to engage the whole moral and religious orientation of the person. Thus the heart, or conscience, is the pivotal expression of the whole person.

In this context Fernando Cardenal's action makes excellent sense. For him the key moral question was that of discerning where his 'heart' was fundamentally focused. Job expresses it simply and directly: 'I take my stand on my integrity, I will not stir. My conscience gives me no cause to blush for my life' (Job 27:6). In the synoptic gospels Jesus simply carries on the teaching of the Old Testament and opposes the formalism of the pharisees. In his teaching the heart is the core source of both good and evil: 'For from the heart come evil intentions . . . These are the things that make us unclean' (Matthew 15:19–20). For Jesus, the whole life of a person must be directed by love: 'You must love the Lord your God with all your heart...You shall love your neighbour as yourself' (Matthew 22:37). Thus in Jesus' teaching the commitment of the heart is the key to moral commitment and to spirituality.

Saint Paul is the first to reflect systematically and theologically on Christian experience. He was aware of the world around him and he borrowed words from that world and transformed them in a Christian context. One of these words was 'syneidesis', to know yourself, to have self-knowledge. Paul uses the word in that sense, but he links it with the biblical idea of heart and conscience: 'The aim of our charge is love that issues from a pure heart and a good conscience and sincere faith' (1 Timothy 1:5). Thus, for Paul, what the Christian requires is a 'pure heart' (that is, genuine integrity), 'sincere faith' (which forms a firm basis for conviction) and a 'good conscience' (that is a conscience that reconciles integrity and faith). Paul often says that his conscience is 'clear' (2 Corinthians 1:12), for the role of conscience is 'to accuse' (Romans 9:1) or 'to acquit' (2 Corinthians 1:12). But this accusation or acquittal is based on and in one's relationship with God. 'I am speaking the truth in Christ; I am not lying; my conscience bears me witness in the Holy Spirit' (Romans 9:1). Conscience is not autonomous; it is subject to the judgement of God. 'True, my conscience does not reproach me at all, but that does not prove that I am acquitted; the Lord alone is my judge' (1 Corinthians 4:4). For Paul, conscience stands in radical relationship to God and is indicative of the stance of the whole person.

It is precisely at this point that Paul discovers the core of Christian freedom. It is not total autonomy; it is freedom from the Jewish law. Paul sees the law as the product of the immature childhood of the chosen people. In Christ, the new chosen people have now reached maturity. The Christian who has entered into a conscientious relationship with God has no need of the law and has transcended and gone beyond the law. 'Where the Spirit of the Lord is, there is freedom' (2 Corinthians 3:17). The mature Christian is free not just of the Jewish law, but of all law.

Today Catholic theology has already reintegrated the biblical vision and it is slowly permeating catechetical instruction. Catholics are thus beginning to rediscover the genuine biblical and historical roots of freedom of conscience and this rediscovery is gradually being applied both within the individual and the communal relationship with God. The radical meaning and primary focus of conscience is the inner spiritual core of the human person as he or she stands before God. This stand should not — and cannot — ignore the wider church community, which is expressed through the church's teaching and belief. But the primary focus will continue its shift from a spirituality of obedience to an asceticism of responsible freedom. Flowing from this must be a renewed emphasis on freedom of conscience in catechetical instruction and preaching. I am not suggesting total moral autonomy, with no regard for other people or the community. In the theological dynamic of Saint Paul, freedom always interacts with obedience to the Spirit of God. For the contemporary Catholic, the Spirit speaks through experience of the self, others and reality itself. This experience exists in a creative tension with the teaching of the bible, the Catholic tradition and the various church magisteria. Thus conscience, truth to one's heart, will gradually supplant obedience as the major touchstone of moral and existential truth.

Part of the problem with obedience in the immediate past has been the almost total identification of all Catholic teaching with the papal magisterium. This is not only untrue to the full Catholic tradition (for such a view only gained ascendancy in the late 19th century), but it is theologically wrong to identify totally the voice of God's Spirit with the papal magisterium. Certainly, the Roman see has been seen as the traditional touchstone of orthodoxy, but never as the sole and exhaustive source of it. The papal magisterium is only one among a number of factors to be taken into account in making a genuinely Catholic moral decision, or discerning a call from God's Spirit to act in a specific way.

John Henry Newman's views on freedom of conscience are well known especially in the context of their relationship to the teaching authority of the pope. His view is expressed clearly in *A Letter to the Duke of Norfolk*. He says: 'Conscience is the aboriginal Vicar of Christ, a prophet in its informations, a monarch in its peremptoriness, a priest in its blessings and anathemas, and, even though the eternal priesthood throughout the Church could cease to be, in it the sacerdotal principle would remain and would have a sway'. His own life is a testimony to his commitment to the primacy of conscience. Newman concludes his discussion of conscience in the book *Certain Difficulties Felt by Anglicans in Catholic Teaching* with the light-hearted but nevertheless telling remark: 'If I am obliged to bring religion into after-dinner toasts (which indeed does not seem quite the thing) I shall drink — to the Pope, if you please — still, to Conscience first, and to the Pope afterwards'. The priority of freedom of conscience was very clear for Newman.

Freedom of conscience, in the truly Catholic sense, involves a radical commitment to discern the will of God through taking into account all the factors that influence one's life. As well as papal teaching, a Catholic must consider the other magisteria — the teaching of the bishops, especially the local episcopal conference, and the teaching of theologians whose traditional authority has been so incorrectly impugned recently by reactionaries. A properly formed Catholic conscience must also take into account the *sensus fidelium*, the belief and practice of the Catholic community. Individual and communal circumstances, as well as the social, cultural and economic situation must be taken into account too. Clearly not all these factors will have equal importance in every moral decision — we act reflexively, not by working through a moral check list. Always, however, the key factor is to seek in radical honesty where the Spirit is leading, so as to be able to stand with integrity before God, ourselves and the community.

Catholics are learning that to dissent from specific papal or church teaching in the moral sphere does not necessarily mean dissent from the church. Most Catholic moral teaching is in the realm of 'ordinary magisterium' which, by very definition, can be and has been changed. Today among Catholics there is a realisation of the historical *relitivity* of so much of church teaching, especially in the moral sphere. There has always been a considerable two-way interaction between Christian moral teaching and prevailing social and ethical norms. In some situations loyal dissent may be a moral obligation, for it is one way in which God's Spirit works in an often unresponsive institutional church.

Freedom of conscience will thus increasingly become the focus of spirituality and action. This will be very important in view of the decline in the present penitential practice of the church. As the priest confessor declines in importance in the formation of the Catholic moral sense, integrity of heart more and more will determine Catholic ethical sensibilities. At times, this will demand a frightening aloneness and great courage as people begin to accept the consequences of their own moral action. Yet, paradoxically, at the same time, the communal nature of conscience is being reasserted. A genuine commitment to Catholic Christianity demands that we recognise that we do not live in a social vacuum. We belong to an interactive system, a network of relationships that is not just local, or even national, but global. Each of us is responsible, with other persons and communities, for the world in which we live.

Catholics cannot abrogate personal and communal responsibility for social justice, peace and care for the environment. Social evil has emerged as the prime sin of our time and has taken over the role that sex played in the moral universe of the last few centuries. The communal Catholic conscience should exercise a critical stance against prevailing cultural, social and economic norms — as can be seen in the prophetic statements of the US bishops on war and peace and the economy, and the draft statement of the Australian bishops on wealth and poverty. As Catholics explore the implications of this, there will be disquiet, uneasiness and, for some, reaction and retreat. It will be an uncomfortable and difficult transition but, despite what reactionaries may claim, there is no way back. The acceptance of the responsibility of freedom before God will be the focus of both the spirituality and action of the Catholicism of the 21st century.

## Tradition

The rediscovery of a genuine freedom of conscience must take place within the context of the ongoing Christian tradition. Some years back, I somewhat naively proposed to a group of Catholic secondary school teachers that it was important to convey to students a sense of tradition. They immediately typed me as a reactionary conservative who wanted to take them back to the dark ages! Their attitude was understandable, for the word 'tradition' in English generally connotes a retreat to the past. Yet it is one of the richest words in the theological lexicon.

Catholicism has always argued that there are two sources of God's revelation — the bible and tradition. The debate about the relationship of the two was one of the most interesting tussles at the Second Vatican Council. [4] There was no debate that the bible is a source of knowledge about God's relationship with humankind; both Protestants and Catholics accept it unequivocally as such. But the role of tradition in this relationship is more complex. After the 16th century Reformation and the rejection of the medieval understanding of tradition by the Protestants, the Council of Trent defined that there are two sources of revelation, scripture and tradition, but it did not spell out the connection between the two. Post-Tridentine Catholic theologians proposed that tradition was a kind of repository of doctrinal teachings the apostles had not managed to fit into the New Testament, such as the doctrine of purgatory or the perpetual virginity of Mary. This somewhat crude understanding has been jettisoned by Catholic theology, and the close interactive link between bible and tradition was reasserted by Vatican II in its decree on revelation, *Dei verbum*. Despite this, there has still not been a great deal of reflection by contemporary Catholic theology on the real meaning and function of tradition. A lot of energy has gone into the rediscovery of the bible over the last 40 years, and tradition has been relegated to the theological too hard basket. Unfortunately, the word has been taken over by ultra-conservatives, such as the schismatic Lefebvrists, who call themselves 'traditionalists'. These people are not traditionalists at all; lacking any real historical perspective, they are unconsciously trying to restore the worship and ethos of French Catholicism of the 19th century and the anti-modernist period at the beginning of this century.

The time has come to say that the bible has been reintegrated into the life of Catholicism, and what we now need to rediscover and to re-emphasise is tradition. Newman makes the interesting and penetrating observation in his book *Via Media* (written in his Anglican period) that Christians derive their faith not from scripture, but from tradition. It seems to me extremely important that Catholics of today avoid the danger of becoming myopically biblical in orientation. Biblicism is a form of fundamentalism. As Newman suggests, a living church must interpret scripture in the light of the living tradition. In certain parts of contemporary Australian Catholicism this seems to me to have been forgotten and a kind of 'protestantised biblicism' reigns supreme (for instance, in sections of the charismatic movement). Without tradition there will be

---

4. For an explanation of this see *Mixed Blessings*, pp. 23–25.

no vital contemporary interpretation of the bible.

It is important to be clear about what tradition is *not*. A genuine Catholic understanding of the theology of tradition does not involve a romantic hankering for the past, a retreat into history to find the solution for today's problems. Above all it is not a nostalgic lingering in attempted reconstructions of the ethos of past epochs, finding there a refuge from the tough issues that press in upon us today. Such hankering after the past is not tradition; it is traditionalism.

Tradition is not a static notion. It refers to a most dynamic aspect of the church, always learning, always growing, always making the most of its opportunities to participate in human reality. Phillip Adams, in a flash of insight, recently described Christianity on Radio National's *Late Night Live* as an 'opportunistic religion'. Christianity must always be able to speak to people of different times and cultures in terms that make sense to *them*. There is a sense in which Christianity, and specifically Catholicism, like any living religion, always adapts itself to its circumstances in a way that could be correctly called 'opportunistic'. This Christian faith is always re-inventing itself. This confronts the contemporary Catholic with a paradox: Australian Catholicism today is not the Christianity of the New Testament, nor the church of medieval Europe, nor even Irish Catholicism of the 19th century. Nevertheless, in a real sense, it is in continuity and contact with all those former manifestations of living Catholic faith. Tradition is the ability to hold these two realities together in a creative tension that inspires the church with memories of the past, sustains it in the present and builds energy and hope for the church of the future.

Christianity is an historical religion; this is its most valuable inheritance from Judaism. Both Jews and Christians have a sense of God's involvement and activity in the flow of life and both value their history as an account of God's interaction with them. So within Christianity — and especially within Catholicism — tradition is both a sense of continuity with the past and the realisation of being a participant in an ongoing process that has not yet finished. The church of today is not beginning *ab initio* — from the beginning — but is a creative force which shaped much of the culture, art, literature, belief and values of the civilization to which we belong. Tradition is a sense of being part of an historical flow, a confidence that you belong to a community that has a past which can be built upon. As an agnostic friend once said to me in an argument: 'You Catholics have had all this time and experience to gradually build up your responses to the basic questions of human existence. I've only got myself'! In the light of this, it is interesting that Australian Catholicism has a remarkably broad and detailed historiography. In the English-speaking world no other local Catholic church has built such a comprehensive knowledge of its past. The field has been so well worked over that it is hard for students to find Australian Catholic topics broad enough to justify major dissertation work! A non-Catholic colleague commented recently: 'The trouble with Australian Catholics is that they are so bloody verbal'.

Tradition, then, is not just a profound sense of history; it is also the perception that, on the basis of this past, the Catholic Christian can look to the future. Tradition is the theology of a changing, developing, evolving church, not one that is statically bogged down in one or other odd period of the past. This is why I have always denied that I am a 'liberal', for I believe deeply in tradition. It is tradition that liberates and frees creative people to live in the present and prepares them to build the future. And they have the confidence to do this because they know that the Spirit of God has worked in the past.

The ironical reality is that the very thing that traditionalists lack is a sense of tradition. Those who have no consciousness of their past will be inexorably trapped in the present, even as they try to deal with the issues of the past. A profound insight of the psychologist C. G. Jung may be helpful here. He says that if we have not passed through the normal stages of human development and have not integrated our past, we will be condemned to go back over and over to face the issues of the past, until we have passed the psychological stages of development that are proper to them. Reactionary Catholics seem to me to be trapped in that very bind; they are always trying to reconstruct an imaginary past, precisely because they have not integrated that past. They seem to seek the unreal chimeras of a past faith that never really existed.

Tradition is vitally important for contemporary Catholics precisely because it places them squarely in the great spiritual, symbolic, historical and cultural flow of the Catholic church. Tradition breaks down the compartmentalising so characteristic of today's world. By giving the Catholic community a sense of place in the flow of Christian reality, it frees them to construct a Christianity that makes sense within the context of the contemporary world and that also looks to the future. All Christian styles are relative to their time and place. Anyone who has done any serious historical work in primary sources knows how different the attitudes, values and beliefs of people of even the recent past are. It is very hard to enter into their world view, to see reality through their eyes and it takes years of study and intimate knowledge of the primary sources to do so. Catholic Christianity has been many things in many historical periods. Contemporary Catholics are challenged to make it something real for the 21st century. This is achieved by finding what is relevant in the tradition today, as well as developing the tradition in new directions.

As Newman says, an image that can help us to make sense out of this is that of a river that is only gradually being explored — but in reverse! Beginning at the source, Christianity is slowly moving downstream, discovering new country, new visions, new attitudes, new insights into the nature of the river itself as it goes. The task of contemporary Catholics is to continue the exploration, knowing that we can build on all that has been discovered so far. Both elements have to be held in tension. So the task before Australian Catholics is not static, but creative and dynamic.

Catholic Christianity has never been, and cannot ever be, merely a religion of static law and unchanging doctrine, established once and for all in the past and true for ever. Newman argues in his *Essay on the Development of Christian Doctrine* that the church's belief and inner life is like an idea which is continually clarified and expanded by development and growth. Newman was never a fundamentalist biblicist, touting the illusion of the purity of the early church as some type of ideal of simplicity to which we should all return. His profound historical sense led him unerringly to the realisation that the more the church grows, develops and changes, the more it becomes truly itself:

It is indeed sometimes said that the stream is clearest near the spring. Whatever use may be fairly made of this image, it does not apply to the history of a philosophy or belief, which on the contrary is more equable, and purer, and stronger, when its bed has become broad, and deep, and full. It necessarily arises out of an existing state of things . . . Its vital element needs disengaging from what is foreign and temporary . . . It remains perhaps for a time quiescent . . . From time to time it makes essays which fail, and are in consequence abandoned. It seems in suspense which way to go; it waivers, and at length strikes out in one definite direction. In time it enters upon strange territory; points of controversy alter; parties rise and fall around it; dangers and hopes appear in new relations; and old principles reappear in new forms. It changes with them in order to remain the same. In a higher world it is otherwise, but here below to live is to change, and to be perfect is to have changed often. [5]

Newman's words, 'to live is to change, and to be perfect is to have changed often', sum up for us what tradition is essentially about: the dynamic of development and change in the life of the church.

To sum up: I believe that the spiritual life of the contemporary Australian Catholic community, and of the Catholic of the coming millenium, should be characterised by an inner freedom that is guided by conscience, and by a profound sense of being part of something greater than the self, that demands a contemporary creativity that is firmly based upon a consciousness of belonging to a great tradition.

## Towards an Australian theology

Contemporary theologians have emphasised that there is no such thing as Catholic Christianity in abstraction. It exists within a given culture. The appalling neologism 'inculturation' has been invented to describe this process. The church must express itself in terms of local culture. If the church fails to achieve this, or even repudiates the local culture (as Rome did, under the influence of the Franciscans and Dominicans, in the 18th century missions in China, when the successful efforts of the Jesuits at inculturation were repudiated), then Catholicism becomes an alien reality. The Swiss Capuchin, Father Walbert Bühlmann, has been one of the most popular exponents of the modern theology of inculturation. [6] (His book *Dreaming about the Church* is a fascinating personal reflection, from a person who knows Rome well, on the contemporary pontificate and the ethnocentrism and opposition of the Roman Curia to inculturation). He argues strongly that the contemporary church is, as Karl Rahner says, a 'world church' in which there is no place for an exclusively European approach to Christianity.

Inculturation means that the church expresses itself in terms of elements from *within* a given culture, elements that give local expression to Christian beliefs and practices. It is true, of course, that

---

5. *An Essay on the Development of Christian Doctrine*, pp. 34–36, 38–40. I have taken the quotation from Ian Ker's *John Henry Newman: A Biography*, Oxford University Press, 1988.
6. See especially his *The Coming of the Third Church* (Saint Paul Publications, Slough, 1974), *God's Chosen People* (Orbis, Maryknoll, NY, 1982) and *Dreaming About the Church: Acts of the Apostles of the 20th Century* (Sheed & Ward, Kansas City, 1987).

the church can also be influenced and, at times, manipulated by the culture in which it exists. European church history is a thousand year long example of this reality. I would argue that this has also happened in Australia; the church is often more 'Australian' than 'Catholic'.

Related to the concept of inculturation is a discussion that has being going on in Australian Catholicism for several years among theologians. It is about what has been called 'Australian theology'. A number of the best theologians in the country have contributed to this discussion, including the editor of *Compass Theology Review*, Father Peter Malone MSC, Father Frank Fletcher MSC, Redemptorist Father Tony Kelly[7] and Father Eugene Stockton of the Parramatta diocese.[8] It was a group of Protestant theologians, however, who first tackled the issue of religion and Australian culture, more than a decade ago. Thinkers such as David Millikan, Robert Banks and Bishop Bruce Wilson sought to find the elements within Australian culture that could be used as vehicles for proclaiming the gospel and relating Christianity to Australian secularism.[9] While their aim was evangelistic, they certainly asked serious questions about the relationship of theology to culture. This was the first attempt by theologians to relate Australian experience to the Christian gospel. In *Mixed Blessings* I expressed considerable ambivalence about both the superficiality and significance of the images used by Wilson and Millikan (ockerism, mateship, Australian humour), and questioned whether these images were able to bear the weight of meaning assigned to them.[10] David Millikan told me himself that he soon found that the paucity of Australian cultural rhetoric inhibited the expression of deeper cultural and spiritual issues and he claimed that the churches usually strongly resisted the theological examination of Australian culture. However, Millikan considers that poets like Les Murray and Bruce Dawe have begun to give some expression to those deeper, inner issues that until now have literally been 'unspeakable' in Australian society.[11] In my opinion, there are a number of other

---

7. For a general treatment of the topic see Malone, Peter (co-ordinator), *Discovering an Australian Theology*, St Paul Publications, Homebush, 1988.
8. Stockton, Eugene, D., *Landmarks: A Spiritual Search in a Southern Land*, Parish Ministry Publications, Eastwood, 1990.
9. See Millikan, David, *The Sunburnt Soul*, ANZEA, Homebush, 1981 and Wilson, Bruce, *Can God Survive in Australia?*, Albatross, Sydney, 1983.
10. *Mixed Blessings*, pp. 207–208.
11. This idea is taken up by Peter Kirkwood in 'Two Australian Poets as Theologians: Les Murray and Bruce Dawe' in Malone, op. cit., pp. 195–216.

writers who have contributed to the articulation of the deeper issues, such as Judith Wright and Patrick White.

The more recent Catholic discussion about Australian theology seems to have ignored the earlier Protestant discussion. While the Catholic dialogue is more explicitly theological than evangelistic in its orientation, it has nevertheless gone back to scratch. I have to admit that I have a feeling of déja vu as I read a number of these Catholic essays in Australian theology. What these writers are attempting is a form of contextual theology, an expression of inculturation. Australian theology, says Peter Malone, is 'a local theology . . . an exploration of our experience and of the doing of theology here'. [12] Tony Kelly assures us that it is not 'nationalist theology', and he goes on to say that

being Australian enters into the foundations of our theological thinking. Either you own it, critically, hopefully, resourcefully; or it owns you . . . For being Australian colours the horizon in which we locate, see and interpret life's mysteries. [13]

Kelly then quotes, with approval, the explanation of Canadian Jesuit theologian, Bernard Lonergan, that the role and task of theology is to mediate between a specific cultural matrix and the significance of religion within that matrix. This explanation reflects Paul Tillich's definition of theology, which forms the basis of the earlier Protestant attempt to do Australian theology in context.

The problem is not the theology, but the cultural context. Kelly honestly admits that a cultural context 'is notoriously difficult to define'. You can say that again! It is hard enough to identify in established cultures like those of Europe, Africa or Asia. But it is especially complex in a migrant society which has enshrined multiculturalism as a kind of cultural ideal. What the theologians lack perhaps is skill and expertise in Australian history and sociology. Several seem to have advanced little beyond Manning Clark's somewhat romanticised 1960s view of the role of religion, and specifically Catholicism, in Australian history.

Even so, the Protestant attempt to describe an Australian theology soon ran into the inability of inherited Australian culture to articulate profound human questions. I do not think that the reason for this

---

12. Malone, op. cit., p. 7.
13. Kelly, in Malone, op. cit., p. 52.

is that Australians are inarticulate; rather it is that the cultural stereotypes of the past have yielded up all that they have to offer and that we stand on the edge of something entirely new. Thus the attempt at articulating a coherent Australian theology seems to me to be premature. A genuine Australian culture is only now beginning to develop; the very fact of a search for an Australian culture indicates that Australia probably does not have one.

Australia since white settlement has always been a migrant society, and competing ethnic groups have struggled for part of the colonial, and then, after federation, the national cake. The bitter sectarianism of the 19th and the first half of the 20th century was a symptom of a tribal form of struggle in which competing ethnic groups attempted to establish themselves and mark out their turf. The Irish, the Scots, the Welsh, the Cornish and even the small groups of Italians and Germans, all expressed their ethnic identity most clearly through their church and their commitment to a specific form of Christianity. All this was presided over by those who had inherited the power of the English establishment and who saw Australia as part of the British Empire. The only questioning of this came from the Irish, based on a deep sense of grievance and the alternative world view provided by Catholicism. This was not effective because, as we have seen, Australian Catholicism wanted to conform and never really drew on its deviant world view to offer a critique of the prevailing Australian ethos. The old multi-ethnic sectarianism has been replaced today by the apparently more civilized and certainly more secularised doctrine of multiculturalism.

However, both the old sectarianism and contemporary multiculturalism come down to the same thing: they are both ways of establishing the rights of ethnic groups to their own identity and to a share in the common wealth of the country, while remaining within the context of a reasonably tolerant and peaceful society. In both the focus is on difference and separate ethnic identity; little attention is paid to what holds the whole community together as 'Australian'.

I think the time has come for a serious critical examination of the doctrine of multiculturalism, and an even more serious reflection on what it means to be Australian. For the post-war continental Europeans migrants and more especially for Asian migrants and other non-Europeans, the old sustaining images — such as the bush, the Anzac legend or the Eureka stockade — are meaningless. How can the bush be meaningful to those post-war Southern European migrants who I know personally and who have never been more than 50 kilometres from the centre of Melbourne, but who have been back to the old country several times? Something more has to to be found to replace the Anglo-Celtic cultural clichés as a bonding force in Australian culture.

Those of us whose families have been here for several generations, have been forced to move beyond the ethnic pretence of being 'Irish' or 'Scottish' or 'Italian' and now have nothing else to fall back on except being 'Australian'. This also applies to many second generation Australians. As a successful television journalist of Greek extraction said to me: 'Good God, I don't want to go back to Greece!' She is happy to make the most of it here, as she does not have — nor does she want — any other option. I would argue that a genuine white Australian culture is only just starting to emerge. It will integrate some elements from the past, but because these reach back to the Anglo-Celtic and European countries of origin and reflect their social, racial, economic and religious divisions, much of this cultural baggage will be jettisoned. And those theologians who look back to the old stereotypes are looking for Australian culture in the wrong place.

Theology is essentially a reflective discipline, and thus looks inward to literature and art for a culture's symbols. But these been interpreted to the point of exhaustion. Australian theology has to strike out in a new direction. I think that Eugene Stockton's book *Landmarks* and Western Australian Baptist pastor Cavan Brown's newly published *Pilgrim through This Barren Land* are beginning to point in the right direction. [14] Both books suggest that a sense of the desert is the real source and foundation of Australian culture. I would broaden this to argue that the entire land is the beginning point and the only foundation for the development of any real culture in this country. The Aborigines have always known this, but they have been treated with such violence and contempt that both European and other migrants have presumed that they had nothing to learn from them. It is significant that some white Australians are just beginning to discover what the Aborigines have always known.

The key to an emerging Australian culture is the unique nature of the place in which we dwell. The constant waves of migrants have drastically disturbed the process of allowing a relationship to the land to develop in the unconscious levels of the white Australian psyche.

---

14. Brown, Cavan, *Pilgrim through This Barren Land*, Albatross, Sutherland, 1991.

I have often thought that Australian theology could find no better place to begin a reflection on non-Aboriginal Australian culture than the marvellous ABC television series and book, *Nature of Australia*. [15] Theology must begin to look outward, to the nature of the place itself, to its geomorphological and natural history, to its land and seascapes and to its unique animals, birds and plants. In this extraordinary environmental context migrant cultures can begin the profound process of conversion from being transplanted Europeans or, more recently, transplanted Asians. There have always been individuals who knew this, but it is only now beginning to emerge more widely.

White Australians are certainly not the first people to define themselves in terms of relationship to a specific place; the Aborigines and virtually all so-called 'primitive' people achieve self-definition in relationship to their environment. However, we are the first community to commence the process of self-definition within the context of contemporary environmental knowledge. The radical implications of this are that we will have to move away from the myth of development at any cost, a myth that still dominates our economic thinking and the policies of successive Australian governments. We will have to abandon the crass materialism and blatant possessiveness that is so characteristic of migrant societies. And it is in this area that the church could become increasingly important. It is just beginning to arrive, late and breathless, on the environmental scene and it is yet to be seen what its significant contribution will be. The development of an Australian theology could well become the bridge between the uniqueness of Australia as a place and the development of an Australian cultural identity. I will return to this question in the final section on environmentalism.

## Social justice

The other key intramural issue is social justice. For Catholics of the near future, sins against social justice will be seen to be as serious as sexual sins were seen to be in the past. I sometimes think it would be good idea if the old moral concept of no 'parvity of matter' were appled to sins against social justice, and that the weapon of

---

15. The whole series is available on two videotapes entitled *Nature of Australia* from the ABC. An accompanying book is also available: Vandenbeld, John, *Nature of Australia: A Portrait of the Island Continent*, Collins Australia/Australian Broadcasting Corporation, Sydney, 1988.

excommunication, used against those who procure abortions, was also used against those whose injustice, exploitation of the structures of society and social sinfulness is equally destructive to the lives of others and especially destructive to the good of the community.

Social justice is one of the areas where the papacy, from the late 19th century onwards, has been a great leader. Pope Leo XIII began the tradition with the encyclical *Rerum novarum* in 1891. This was continued through the century right up to Pope John Paul II's latest encyclical *Centesimus annus* (1991). Development can be seen in the shift from Leo XIII's emphasis on the right of workers to share in the wealth that flowed from the land and control over the means of production, to John Paul's emphasis on the enhancement of human dignity through sharing in the ownership of knowledge, technology and science, the basis of a nation's wealth today.

In Australia social justice has been one of the areas where Catholics since the 1960s have by and large preferred the mediocrity of the attainment of middle class security to the bracing papal critique of contemporary western values.

Some years back I was asked to prepare a paper for a seminar entitled *Sydney 2001 — for People or Profit*. I felt that I had been offered an extraordinarily narrow choice; if these were the only choices, I would choose neither. Both these options play into two of the opposed, but nevertheless pet dogmatisms of our age:

- Profit = capitalism — which today, in an extraordinary misnomer, calls itself 'economic rationalism'!

- People = personalism — the often environmentally destructive anthropocentric view of human existence.

It may seem surprising that a Catholic is apparently criticising 'personalism'; but hold your fire on this, I will come back to it when dealing with environmental issues. Let us tackle profit first!

## Profit

At the time I was writing the paper on *Sydney 2001*, I heard the prize-winning Radio National *Encounter* programme featuring Dr Charles Elliott. He comes from the English Evangelical tradition and was an economic adviser to the last British Labour government. As he pointed out in the programme, 'economic rationalism' is a euphemism for free market capitalism. It poses as a science and purports to examine the forces that underpin the so-called 'free market' which still dominates so much of our economic activity. In reality, these free market forces are governed by a form of determinism that is about as scientific and predictable as fate. Economic rationalism is nothing more than an 'open slather' for competitiveness, individualism and particularism to run riot, allowing the most ruthless and amoral to enhance their financial power and prestige at the expense of others, especially by the manipulation of debt. In more traditional days, the Catholic church used to call such activity 'usury' and viewed it as a more important mortal sin than sexual immorality. It would not be such a bad thing if that understanding were renewed today. It is ironic that a number of Australia's recent free-market buccaneers are Catholics and that it was state and federal Labor governments that created the conditions for this totally unproductive and immoral form of capitalism.

The Letter to the Ephesians speaks about the human struggle against 'the sovereignties, principalities and powers, the rulers of the present darkness' (Ephesians 6:12; see also Colossians 2:15). These may seem like angelic entities, but Saint Paul understands them to be the incarnation of the destructive forces that are loose in the world. The nature of these forces varies from period to period; Charles Elliott sees them today as the forces that make up free market capitalism. In Australia we have learned to our cost that the New Testament is right: behind such forces lies a demonic and destructive reality.

The Catholic church has a long and coherent critique of the ethics of profit. The roots of this lie in the medieval period when the church integrated into its tradition an understanding of society as a corporate reality, in which, as members of the human community, we are all interrelated and have a responsibility for each other. This in turn is founded on the biblical doctrines of co-responsibility, love for others and distributive justice and equity, whereby a person's worth arises from their intrinsic humanity and their value before God, and not from their human wealth and power. Flowing from this is the sense that all have a right to share fairly in the goods of the earth. In the Hebrew scriptures the prophets reserve some of their worst vitriol for those who 'corner the market' and destroy other economically weaker people in the process.

In the 20th century church, the popes from Pius XII to John Paul II have mounted a major and coherent critique of free market capitalism. A statement of Paul VI in the encyclical *Populorum progressio* (1967) could well be applied to the rampant capitalism of Australia in the 1980s. It was unfortunate, the pope said, that an economic system has been constructed

which considers profit as the key motive for economic progress, competition as the supreme law of economics and private ownership of the means of production as an absolute right that has no limits and carries no corresponding social obligation. This unchecked liberalism leads to dictatorship rightly denounced by Pius XI as producing 'the international imperialism of money'. [16]

Pope John Paul II's two most recent encyclicals *Sollicitudo rei socialis* (1988) and *Centesimus annus* (1991) also have many critical things to say about capitalism. Perhaps the time has come for the church in Australia to start to use some of its ecclesiastical weapons — such as naming names and even excommunication — against Catholic entrepreneureal criminals and the governments that have encouraged, by their policies, the corruption of our economic, civic and social life.
    Some of this has been taken up in the Australian bishops draft statement *Common Wealth and Common Good*. The statement has had, of course, its critics, one of whom, philosophy professor Lachlan Chipman, is quoted as saying that the statement presents 'a Marxist, liberation theology principle that some are rich, some are poor and money must be taken from the rich and given to others', and that it is 'as much Marxist in tone as Christian'. [17] Only a person who does not know the tradition of Catholic social teaching could make such a silly statement. It mirrors the statement of Gerard Henderson that the pope's encyclical *Sollicitudo rei socialis* 'touts fashionable left wing orthodoxy' and that it should 'be withdrawn and rewritten'. [18] There is absolutely no doubt that the popes have condemned marxist socialism; the best a critic can say is that they have been ambivalent about capitalism. But that hardly gives it a ringing endorsement. In *Centesimus annus* John Paul II warns western capitalism of the risk of seeing the collapse of marxism in eastern Europe 'as a one-sided victory of their own economic system'. Even within the context of marxism's collapse, the pope says clearly:

While acknowledging the legitimacy of private ownership of the means of production and the value of a free market economy, the church insists that social life must be governed not by the dictates of the market alone,

---

16. Pope Paul VI, *Populorum progressio*, par. 26.
17. *The Australian*, 30 April 1991.
18. *The Weekend Australian*, 16 April 1988.

but first and foremost by concern for each human person and by the desire to create a human community marked by co-operation with a view to the common good, mindful that the earth's resources are destined for the use of us all. The church views labour and all economic activity within the higher perspective of mankind's transcendental vocation.

A debate about the stance of Pope John Paul toward capitalism in *Centesimus annus* has also arisen in the United States. It began between two Catholic theorists about the compatibility of different types of capitalism with the encyclical's teaching. The 'neo-conservatives', championed by recent convert (from Lutheranism), Richard John Neuhaus, argue that the pope has affirmed the 'new capitalism' (by which he really means old style American capitalism of a totally free market, unfettered by any government regulation). Michael Novak, the author of *The Spirit of Democratic Capitalism*, argues that Pope John Paul supports his contemporary free market style capitalism based on the theory of economic liberty. (He points out that his book was translated into Polish in 1984 and that the pope got a copy). From the mainstream Catholic church, Jesuit Father David Hollenbach of Weston School of Theology argues that *Centesimus annus* warns against the 'idolatry of the market' and adds that 'the pope is saying yes to a free-market economy, but with a significant set of "buts". The "buts" have to do with the inclusion of large numbers of marginalised people, people who have been left out' — and here Hollenbach refers to the homeless and to the eleven million people looking for work in the US. [19]

The Catholic experience in Australia over the last thirty years has been that the papal discussion of social justice has remained at the theoretical level and has not really impacted on the lives of ordinary Australian Catholics. The reason for this seems obvious enough: any debate tends to remain at the level of the theories of capitalism and socialism. There has been little attempt to translate papal teaching into a context meaningful for ordinary Australians, which is precisely what the Latin Americans have done. It is good that *Common Wealth and Common Good* has begun this local translation.

What does the papal teaching mean at the human level and what will it mean in the future? The answer lies in a return to the radical teaching of the New Testament about wealth and security; it is only in this context that Catholics of the coming decade are likely to find

---

19. Quoted in the *National Catholic Reporter*, 17 May 1991, p. 3.

their inspiration and to begin to integrate papal teaching.

Let me illustrate what I mean. People who know Sydney well will know Belmore Park. It is right beside Central Railway Station in a now rather unfashionable part of town. It is a hangout for drunks and homeless men, and it can be a dangerous place after dark. Some time ago I saw a group of homeless men sitting in the park; there were about ten of them, a mixture of whites and Aborigines. They had a couple of bottles of cheap sherry which they were handing around. There was absolutely no consciousness of race and no question that the wine should not be shared among them all. I am constantly astonished at the readiness of the long-term homeless to share whatever they have with their fellows. It amazes me that people who have next-to-nothing seem so ready to share it. It is often very different for those of us who do have security. My impression is that the more you have, the less you are willing to give to others, unless, of course, it reduces your taxes or enhances your reputation — or both! We live in an age utterly preoccupied with security, with 'nest eggs', as those traders in human insecurity, the insurance companies, call them. My own experience is that the older you become, the more the temptation grows to be concerned with security. Some older people can be incredibly selfish, the product, presumably, of a lifetime of small decisions not to give to others or to share with them.

For me the core of the New Testament teaching about wealth lies in its approach to security. Jesus calls his disciples 'to leave all and follow him'. This is directed to all of his followers, not just to an elite. What does it mean? Jesus certainly does not counsel destitution, and the beatitude 'Happy are the poor' does not say that starvation is a desirable state. What Jesus calls his followers to is what we might call a 'risk-taking flexibility' that involves the disciples moving beyond the parameters of a secure existence based on an inflexible and stable life-style, to an ability to live lightly on the earth, with a constant openness to new possibilities. But this can never be achieved by people who are so tied down by the need for security that they are turned in on and concerned only about themselves. It is within this context that Jesus' statement about the flowers of the field and not worrying about the future makes sense (Matthew 6:25–34). Genuine followers of Jesus should have a flexibility that allows them to be always ready to trust God sufficiently to leave all-embracing security behind.

The rich young man of Matthew's gospel (19:16-30) is a paradigm of the person who cannot hear Jesus' teaching about security and abandonment, 'for he had great wealth'. This striking story is followed by a discussion of the consequences of wealth: 'How hard it will be for those who have riches to enter the kingdom of heaven'. The disciples were amazed at Jesus' words, for, no doubt, they too shared the traditional Jewish conviction that riches were a sign of blessing from God. This notion of wealth as blessing has been carried over into fundamentalist Protestantism which, as a result, is socially conservative. Such people conveniently forget the Old Testament teaching about the corporate nature of society and of our interlocking responsibility for each other. They also forget the prophetic teaching about the need to share the goods of the earth on the basis of equity and social justice.

Jesus repeats his teaching about wealth in even stronger terms by saying that it would be easier for a camel to pass through the eye of a needle than for the rich to enter heaven. His followers, in the translation of the *Revised Standard Version*, were 'exceedingly astonished' (Mark 10:27). The passage makes it clear that Jesus' approach to wealth and security was radical and revolutionary. It gives no comfort to wealthy Catholics and it throws the onus onto them to demonstrate their commitment to social justice. It is ironic that, having left the working class behind, Australian Catholicism has become so tied to a bourgeois scale of values. Yet, paradoxically, because the middle class is usually the most conscientious group in our society, it is also most likely that it is from this same middle class that those who take Jesus' teaching seriously will emerge. The poor are too concerned with the struggle for survival, and the rich too ruthlessly amoral, to be concerned with the articulation of the serious questions of conscience regarding social justice.

## Living a commitment

It is easy for prosperous Catholics to conclude that Jesus' teaching about wealth and security is only directed to the most radical Christians, perhaps only to members of religious orders. I do not think that this is true. However, because religious orders in Australia give their members such total security (from guaranteed board and lodging to an adequate health cover), it is easy for them to lose sight of the fact that poverty is about flexibility and a willingness to step beyond security. This is not to say that there are not many religious in Australia who have not committed themselves radically to social justice. The recent murder of Australian Josephite Sister, Irene McCormack, in Peru, reminds us that there are many religious men and women risking their lives in the service of justice. Members of order have the great advantage of committing themselves totally and in a permanent context to work for others. I am often tempted to ask reactionary Catholics, who are so quick to criticise contemporary priests and religious, how many martyrs the reactionaries have given to the church recently. It is significant that the three American sisters (and the lay woman Jean Donovan) killed in El Salvador, the murdered Jesuits of the University of Central America and now Sister Irene McCormack, were all contemporary religious, working for justice and the poor.

It is my view that with the continuing decline in the number of religious in the traditional sense, those most likely to try to live a radical ideal of commitment to the God of justice will be younger single and married people and older married women whose families have grown up. It is from this group that the 'religious' of the future — those totally and radically committed — will emerge.

So far in this discussion about security, I have not taken into account the position of families with children and the deep sense of responsibility that people in committed relationship have to each other. I often feel that it is married people with children who are those who give most witness to the insecurity and flexibility that the gospel teaching demands. If one or other partner is sacked or made redundant (a real possibility these days), a destabilisation of family life can occur. I believe that many married couples with children in Australia are living a profoundly committed life. Many of them have had to learn to trust God deeply. Conversely, marriage is also the period when other people begin the process of building a security base that is gilt-edged and, consequently, spiritually crippling.

But people are not always engaged in the care and education of their children. Before and especially after the years devoted to family and nurture, there are long periods when people have time, talent and opportunity to experiment with the gospel call. How they will do it will vary with their age, experience, personality and opportunities.

Let us start with young people. While I deplore the anti-Catholic and proselytising tendencies (especially in Latin America) of evangelical organisations such as Youth with a Mission, at the same time I admire their ability to harness the enthusiasm of young people and to direct it towards ministry in a coherent setting. My experience of the young people attract such organisations is that they are generous idealists, who in a Catholic setting in another age would have been candidates for religious life. The generosity of the young is quite extraordinary, but it needs direction and formation. Unfortunately, within worldwide Catholicism, it is the reactionary organisations like Opus Dei, Communion and Liberation and the Neo-Catechumenate who have been very successful in harnessing the idealism and energy of young people, and to an extent this is also true of Australia. Mainstream Catholicism needs to develop more organisations like the Jesuit Volunteers or small local communities that are committed to social justice and to ministry within the world, without necessarily being committed to permanency or celibacy.

There are also younger married people who live a deep level of commitment to justice, often while caring for young children. The ABC programme *Encounter* featured one such couple in the episode 'A Family Life'. It tells the story of a husband and wife who, while caring for three children, one of whom is permanently incapacited as a result of a road accident, also look after several homeless people in their own home, as well as fulfilling a number of other commitments to people in need. The extraordinary generosity of such people is quite remarkable and demands a high level of spirituality and very considerable trust in God. What I think will continue to happen in the Australian church is that young people desiring radical commitment will no longer move toward traditional religious life but will, in fact, achieve remarkable levels of involvement while living in the world. They will do this either as single people or as people living in a mutually supporting relationship. Because they are outside the context of insititutional religious life, they will be often faced with real insecurity and will be living closer to the gospel ideal. In this way they will be sharing the ideal of radical commitment that first inspired religious orders.

The elderly can be preoccupied with security; but they can also be extraordinarily free, generous and persevering in their commitment to social justice. As the numbers in this age group in Australia increase, it will become more and more important for the church to benefit from their experience, sense of reality, tolerance and maturity.

It is through such committed people of whatever age that the Australian church can begin to articulate applied social teaching. It is at the grass-roots level that social justice begins to impinge on the structures of society. The major contribution of the development of *Common Wealth and Common Good* was the chance that it gave for groups and individuals in the community, who are already working for social justice to articulate publically their reflections on what is happening at the grass-roots. Through reflection on this experience attitudes will emerge that will give shape and meaning to the more abstract and 'universal' teachings of the popes.

## The church and Aborigines

Another area of ministry that has changed in Australia over the last two decades is pastoral work among Aborigines. The most perceptive piece that I have read on this is Father Frank Fletcher's *Good News for the Aborigines*, a talk first given in 1989 in the Great Synagogue in Sydney. [20] Frank Fletcher points out that while he works in the Aboriginal Catholic Ministry in Sydney, the real ministry is carried out *by* Aborigines *to* Aborigines, with white priests, sisters and laity working in solidarity and support. This is a radical move away from the paternalistic attitude which presumed that ministry was whites 'looking after' Aborigines, who were supposed to be unable to do this for themselves.

Fletcher makes several major points in this paper, but the core of his argument is expressed in his first point: the result of standing with Aboriginal people is 'the shock of difference and isolation'. He highlights the fact that Aboriginal attitudes to the world and their religious views are totally alien to Europeans and vice versa. As Fletcher says:

I do not know, from the inside, how Aborigines feel. I am an outsider, what they call a 'gubbah'. With every day I become more convinced that Aboriginal people are different. Meeting their differences makes me aware of the narrowness of my own consciousness. I carry much of the European history of the church along with the viewpoint of middle class Australian society. My Aboriginal friends have spent their life-time at the bottom or on the fringes of Australian society and they carry a history of manipulation and patronisation, some of it at the hands of church agencies and institutions. It is not only a different history we both carry, but a separate mentality.

Writing a critique of the articles on Aboriginal religion in Mircea Eliade's new *Encyclopedia of Religion*, Tony Swain of Sydney University emphasises that until the arrival of the Europeans, Aboriginal cultures 'were localised and introspective, cherishing the significance of beliefs and practices belonging to a small group of people, often only a handful'. [21] In other words, one of the most

---

20. Unpublished talk given at the Great Synagogue of Sydney, 7 December 1989.
21. Swain, Tony, 'Belonging to the Emperor: An Australian Perspective on the Encyclopaedia of Religion', *Australian Religion Studies Review*, 2/3, p. 92.

profound differences about Aborigines is that it is impossible to generalise about their culture and their religion, which differs markedly from place to place. This emphasis on difference should make Australian Catholics wary of glib comments about 'Aboriginal spirituality', as though this were something reduceable to the categories of our theology. 'Spirituality' is a western term and it conveys a specific meaning content. The easy adscription of the term to the beliefs of the Aborigines is not only a distortion, but a patronising refusal to see those beliefs within their own specific context. Fletcher argues that it is our separation from and lack of understanding of the Aborigines that leads to our 'mentality of superiority and patronage'. He says:

Aborigines stubbornly remain different. They are not going to become European clones whether in society or in religion. That inward stubbornness, preserving their own selves to themselves, that is what, before anything else, we have to respect about their struggle.

Fletcher argues that this means that other Australians should not be 'pushing into the privacy of Aboriginal identity and spirituality'. Rather we should be supporting Aborigines in their struggle for identity, land and education. Recently the Australian Catholic bishops issued a pastoral letter (2 September 1990) on the need for 'a just and proper settlement' between white and black Australians. The pastoral sets out the requirements for this:

1. A secure land base for dispossessed Aboriginal communities with special attention being given to traditional communities on their own lands.
2. A just process for the resolution of conflicting claims to land and its use.
3. An assured place for powerless Aborigines in our political processes.
4. A guaranteed future for Aboriginal culture and tradition with legal protection of Aboriginal heritage and public education of all Australians about Aboriginal history.

The churches generally and the Catholic church specifically have improved in their attitudes to Aborigines, as the pastoral letter shows. But Fletcher also points to the failure of the churches here, a failure that reflects, he says, their 'European arrogance': 'Even today we have vocal fundamentalist churches among the Aborigines who regard their whole culture and religion as pagan and useless'. Fletcher maintains that white Christians must respect 'the eruption of feeling' among Aborigines for their own religion and culture. As a result of this many Aborigines are turning back to their own beliefs. Consequently, 'It is only a tiny minority who have the desire to dialogue with us about the divine mystery'. If they do come to Christianity, it must be in their own way, for in the past Christianity has come to them with the genocidal conqueror. Fletcher even risked comparing the treatment of the Aborigines with the Jewish holocaust (a rather brave thing to do in the Great Synagogue!).

Fletcher's paper also points to a continuing problem: the unwillingness of white society to be 'confronted before God' about the theft of the land of Australia by our forebears and the consequent injustice to the Aborigines. Most whites assume that

Whatever we possess must not be questioned. The competitive people of today will not be burdened by accusation over the the actions of rapacious ancestors. The law of the market place, they say, demanded that unfit people be replaced. Those who remember and feel for the Aborigines . . . are simply feeble minded. Most Australians remain assimilationist in attitude.

That final statement has been backed up very strongly by the findings of the Royal Commission on Black Deaths in Custody. The appalling attitude of most white police simply reflects the equally appalling attitude of the vast majority of white people in Australia.

The church is one place where there is a chance that attitudes can be changed and the need for repentance for personal sin and those of our ancestors can be confronted. The task is there for the Catholic church of the next few decades. It remains to be seen if Catholics have the moral fortitude to tackle it.

Talking about Aborigines and their relationship to the land, is a good way of introducing the final issue of this book: Catholicism and ecology. Aborigines see the land as a living reality. For them it is not an inert resource that will provide $x$ cubic metres of timber or tonnes of coal. They look at the land with reverence and not with the greedy eyes of mining or logging companies or developers. Recently we had the spectacle of the chairman of a large mining company (and his economic acolytes) telling us that the Aboriginal commitment to the land — and specifically to Coronation Hill in the Northern Territory — is nothing more than 'superstition'. One might be justified in asking if Australian mining interests were not engaged in the equally superstitious worship of Mammon.

# *The church looks outward*

The key problem facing the Australian church in this decade and in the first part of the coming millenium is the ecological crisis. All other theological, moral and pastoral issues fade into insignificance beside this. For the Catholic church to stand idly by while Australia is environmentally ravaged would be a denial of its deepest responsibilities. The environmental crisis cannot be examined in isolation from what is going on in the rest of the world, so I will first try to provide some context for the discussion.

## The environmental crisis

Over the last 200 years two competing ways of understanding the world have faced each other across a seemingly unbridgeable gap: the scientific view of life and the religious view of life. However, the unbridgeable gap has been getting narrower lately, and at least a one pontoon — it is probably not yet a bridge — has begun to connect both sides. The one issue that is bringing the two together as never before is the environment. Of course there are still entrenched positions: at one extreme are the fundamentalists who tout 'creation science', and at the other are the materialists who despise religion and spirituality as primitive and unhelpful approaches to human existence. There is still a lot of suspicion on both sides. But the environmental pontoon is nevertheless carrying increasingly heavy two-way traffic.

The enormous problems now facing our planet are bringing the two sides together. These problems are well known. The greenhouse effect, created by the burning of fossil fuels, could wreak havoc with agriculture and with sea levels, which could rise significantly. The recent cyclone in Bangladesh has already shown how much of the land in the delta of the Ganges — as in many of the small Pacific nations — could be lost by just a slight rise in sea level. The depletion of the ozone layer especially threatens southern hemisphere countries like Australia. So much has already been lost. By the year 2000, twenty per cent of the earth's animals, birds and reptiles could vanish in an unprecedented wave of extinction. Many species are on the edge: the black rhinoceros is one example. Ancestors of this animal have lived on the African savannahs for fifty million years, but between 1970 and 1986 poachers have massacred the population from 65 000 to less than 4500. The rhino's horn is believed by oriental medicine to have aphrodisiacal and medicinal properties and it also provides material for dagger handles in North Yemen: a whole ancient species is brought to the edge of extinction, not by human need, but by human cupidity and selfishness.

Australia has a very poor record in animal survival since 1788. Vincent Serventy (one of the great pioneers of Australian conservation) noted in 1966 that forty-two per cent of Australian marsupials were either extinct or rare. [22] As I was writing this section there was an item on ABC television news about the mallee fowl which is on the edge of extinction right now. The reason for such extinctions is the introduction of feral animals such as foxes, wild pigs and cats, the environmental destructivness of other introduced species such as rabbits and sheep, and above all the alteration and destruction of habitat by human intervention. European civilization has been extraordinarily destructive in this country. The Aborigines also seem to have had destructive effects upon the Australian environment. The head of the department of mammals at the Australian Museum, Dr Tim Flannery, notes that 'Australia once supported a diverse fauna of large animals . . . but all of them vanished between 40 000 and 20 000 years ago', a time that probably coincides with the arrival of the Aborigines. Flannery notes that

Australia then settled into a period of ecological quiescence, and it was not until after 1788 that the extinctions started again. This time, with European exploitation, a further twenty mammals and three birds (so far) have vanished along with many plants and uncounted insects. [23]

The environmental destruction that has been wreaked in Australia so far by human beings is massive.

The primary cause of this destruction is the disappearance of creatures' natural habitats; a secondary cause is one less well known, but even more directly destructive. The ever-growing human population is making increasing demands for fauna and wildlife products:

The combined effects of subsistence and commercial hunting, sport and capture for pet, zoo and research trades have been directly responsible for the loss of at least 15 mammals and 20 bird species (worldwide). Currently,

---

22. Serventy, Vincent, *A Continent in Danger*, Andre Deutsch, London, 1966, p. 212.
23. Flannery, Tim, 'Plague in the Pacific' in *Australian Natural History*, 23/1(1989), pp. 23, 24.

for at least 20 per cent of all birds considered to be endangered, trade is the major threat. For mammals the figure is about 31 per cent and for reptiles the figure climbs to over 50 per cent. [24]

Despite laws and the efforts of Customs, Australian birds and reptiles are seriously threatened by illegal trade. Japan and, to a lesser extent South Korea and other Asian nations, are the major consumers of the trade in endangered species and Singapore is its principal laundering point. Japan and Taiwan are also notorious for their destruction of the life systems of the Pacific Ocean through drift net fishing (the 'wall of death', as it is more accurately called) and Japan is specifically responsible for the near extinction of several species of whales through commercial exploitation.

The arrogance and selfishness of those involved in such activities highlight the fact that we face an underlying problem of incalculable importance. It can be simply stated: it is the question of the relationship of human beings to the rest of the created world. Most human beings see themselves as the centre of reality. We assume that the world exists for us and that the whole of cosmic reality is focused on us. Our consciousness is dominated by what American Catholic thinker Thomas Berry calls 'exploitative anthropocentrism'. By this he means that human beings have taken themselves as the focus, norm and final arbiter of all of the rest of creation. We assume that the earth exists simply for us and that its value is derived totally from us; we are the measure of all reality.

When this is questioned most people respond: Why not? As human beings it is natural for us to see things from our own perspective. Indeed, everything is species-centered. But there is one difference between us and all other species. Our consciousness is developed to the extent that we can see reality as a whole and from points of view other than our own. We have the faculty of imagination; we can see other possibilities. It is up to us to use this imaginative, inventive faculty on behalf of all other species, on behalf of the whole of creation.

Rain forests are being destroyed at the rate of the area of England every year in South America, West Africa, the Philippines, Thailand, Burma, Malaysia, Papua New Guinea and Australia. Australia has already lost 75 per cent of its rainforest since 1788. By 2020 — not

---

24. Milliken, Tom, *Decimation of World Wildlife: Japan as Number One*, Penang, Malaysia, 1988.

so far into the future — all the great world forests will be gone, except those in the West Amazon and West Africa. The whole diversity of life, which has taken billions of years to establish, is threatened with extinction in a miniscule period of 50 years.

The earth shows the signs of our barbarism. It is estimated that we have ten to twelve years to do something about the ozone layer. Current droughts and resultant starvation in Ethiopia, floods in Sudan where the Blue and White Niles meet, acid rain and the biological death of thousands of lakes in North America, the destruction of the Black Forest in Germany and forests and lakes all over Eastern Europe, the leaching and salination of Australian soil (the New South Wales government has already conceeded that 70 per cent of the land surface of the state is degraded), and the pollution of the Murray-Darling river system, are all the results of human decision and human action. And I have not even mentioned the greatest problem: human over-population, which is responsible for so much destruction of the earth.

## 'To conquer the earth'

Human beings are the cause of all of this; the degradation of the earth does not 'just happen'. Most of us, however, are afraid to use our imaginations to see ourselves from the perspective of the rest of creation. If we did we might begin to realise that we are not

the splendour of creation but . . . the most pernicious mode of earthly being. We are the termination, not the fulfillment of the earth process . . . We are the affliction of the world, its demonic presence. We are the violation of the earth's most sacred aspects. [25]

These are the words of Thomas Berry, the American cultural historian and theologian who describes himself as a 'geologian'. His thought is radical and far-reaching and yet strongly in the Catholic tradition. His influence on much that I say about environmental issues will be obvious. He constantly reflects on the relationship of humankind to the world and the rest of creation, and he argues that ecology must become the central issue facing the Catholic church (and every other church).

We today are seeing the consequences of the sins of our forebears. The industrial society that we have inherited was built on the

---

25. Berry, Thomas, 'The Dream of the Future: Our Way into the Future', *Cross Currents*, summer/fall 1987, p. 210.

myth of progress, a notion that came into its own in the liberal culture of the mid-19th century. The myth was that the historical forces that had carried European culture towards an ever-fuller life, based on the presumed super-abundance of the earth, would continue. Science and technology would lead European civilization toward material happiness and perfection. This dream of limitless resources inspired our white Australian forebears to 'conquer the wilderness' (and massacre the Aborigines, tear down seventy-five per cent of the rainforests and destroy much of the flora and fauna as they went). Something of this spirit was amusingly expressed in the 1888 centennial cantata composed especially for the occasion by the Melbourne Congregationalist minister, the Reverend William Allen. It was performed in the Melbourne Exhibition building for the centenary of the foundation of Australia. This is how *Australians 1888* describes it:

Allen's cantata was dedicated to the theme of material success. From 'a land by civilization's step untrod' Australia had been transformed by the 'pastoral pioneers' and the 'magic of gold' to a nation of 'myriad-peopled cities'.

> Where the warrigal whimpered and bayed
> Where the feet of the dark hunter strayed
> See the wealth of the world is arrayed
> Where the spotted snake crawled by the stream
> See the spires of a great city gleam
> Is it all but the dream of a dream? [26]

This dream still dominates contemporary economic thinking. Thomas Berry says that the development myth 'has fostered in our western tradition a profound resentment against our human condition'.

Economic disfunction is usually expressed as deficit expenditure; income does not balance outflow. The federal treasurer is constantly telling us about Australia's balance of payments problem, created by our penchant for expensive imports. What has never been calculated in our desire for development is the deficit that the earth and its resources suffer in this exchange. Thomas Berry says that this is

the deficit involved in the closing down of the basic life-systems of the planet through the abuse of the air, the soil, the water and vegetation . . .

---

26. Quoted in Davison, Graeme, McCarty, J.W. & McLeary, Ailsa (eds), *Australian 1888*, Fairfax, Syme and Weldon, Sydney, 1987, pp. 23–24.

recently both textbook economics and corporation practice have ignored the implications of such data or have given it minimal attention . . . We can be sure that whatever fictions exist in Wall Street bookkeeping, the earth is a faithful scribe, a faultless calculator, a superb bookkeeper [and] we will be held responsble for every bit of our economic folly. [27]

The environmental deficit is an ultimate one: once the topsoil is gone, it has gone; once the rainforest is gone, it has gone; once the animal and plant species are extinct, they are extinct. Nothing will bring them back. We face a problem never faced before in history. The industrial economy upon which capitalism is based is closing down the planet as a habitable place.

Many still refuse to see that the myth of development is fatally flawed. They believe our problems will eventually be solved. Some argue that this will be achieved, for example, by a colony in space. Even if we allow that this is achievable in the long-term, the question still remains as to what we are going to do in the meantime? Also it reveals an extraordinary callousness toward the earth, which has nurtured us for so long; it is simply seen as disposable, once it has outlasted its usefulness. This is an extreme example of the mentality that the earth itself is merely a resource and that all resources are disposable. Writers like Julian Simon (in *The Ultimate Resource* (1982) and *The Resourceful Earth* (1984)) still say that there are no limits to growth. Simon argues that every generation in modern times has lived better than the previous generation, that we must not lose our nerve, for science and technology can resolve all our difficulties. This highlights the way that development economics is driven on by blind faith in the myth of progress, even though the dire consequences of unending growth are so well documented. Development economics has not, as Berry says, created 'wonder world', but 'wasteworld'.

The prophets of the environmental movement have come from outside the church; the Christian response so far has been largely confined to important individuals. The key figure in the development of an early Christian response was ironically an American medieval historian, Lyn White. In 1967 he published in the journal *Science* the essay 'The Historical Roots of Our Ecological Crisis'. White conceded that it was science and technology that had provided the tools of ecological destruction, but he argued that it was the Judeo-Christian tradition that provided the theoretical framework for the

---

27. Lonergan, Anne and Richards, Caroline (eds), *Thomas Berry and the New Cosmology*, Twenty-Third Publications, Mystic, Conn., pp. 8–10.

exploitation of nature. At the heart of the problem was the exaltation of humankind to the mastery of creation. 'Christianity is the most anthropocentric religion the world has seen'. White claimed that the creation story in Genesis vitiated the relationship between humans and nature and that nothing would change 'until we reject the Christian axiom that nature has no reason for existence save to serve man'. His article caused an enormous stir and much of the effort of more ecologically minded Protestant theologians since has been directed toward dealing with his arguments.

Roderick Nash's book *The Rights of Nature* clearly demonstrates that Protestant thinkers in the United States have been especially active. [28] The process theologian, John B. Cobb, has been another key figure in the development of environmental ethics. Cobb argues that contemporary Christianity is inadequate to deal with the environmental crisis and that morality must be expanded to include the concept of the rights of nature. He argues that humankind is the highest point of a pyramid of life in which all reality is valuable to God and thus sacred. Animals have rights to their own existence and must be granted the possibility of happiness and not exploited by humans. Australian biologist and theologian, Charles Birch, has co-authored with Cobb *The Liberation of Life*, which argues that all forms of life have a potential for an experience that must be respected by humans. In other words, ethics is not confined to humankind but extends to all forms of life. [29] Birch, the retired Challis Professor of Biology at Sydney University, is a religious thinker of great originality and is a winner of the Templeton Prize for progress in religion.

Among the Catholics, the great pioneer is the French Jesuit palaeontologist, Pierre Teilhard de Chardin. He never spoke directly about ecology, but his image of the totality of nature moving toward 'omega point', the summit of human development and consciousness, finally achieved only in Christ, indicates a sense of the continuity and interconnectedness of all reality. Thomas Berry is the most radical contemporary Catholic thinker to apply Teilhard's thought to environmental issues. He is something of the Teilhard de Chardin of the contemporary Catholic church. Most of his work has been published in articles in the periodical *Cross Currents*; he only has

---

28. Nash, Roderick, *The Rights of Nature: A History of Environmental Ethics*, Primavera Press/Wilderness Society, Leichhardt, 1990, pp. 87–120.
29. Cobb, John B. & Birch, Charles, *Liberation of Life: From the Cell to the Community*, Cambridge University Press, 1981.

one published book, *The Dream of the Earth*. [30] Berry's ideas have been further developed by the physicist, Brian Swimme. Another major Catholic ecological thinker (who often visits Australia to teach) is the Irish Columban priest, Sean McDonagh; his environmental and theological writings are eminently practical. He worked for ten years with the T'boli people in south eastern Mindanao in the Philippines, and he has witnessed the total destruction of their culture and way of life as their rainforest was destroyed over a period of twenty years by the timber companies. McDonagh highlights the interconnectedness of the environmental destruction in the Philippines: the loss of the forests has led to enormous soil erosion which silts up the rivers and flows into the sea where it is now beginning to destroy the coral reefs. His two books *To Care for the Earth* and *The Greening of the Church* have had considerable influence both inside and outside the church. [31]

## Origins of today's cosmology

Thomas Berry reminds us that a key factor in our attitude to ecological issues is what philosophy calls 'cosmology': our understanding of and attitude toward nature and the universe. Contemporary western cosmology has been formed from several diverse sources, the most important of which is, as Lyn White has pointed out, the creation account in Genesis. Western culture has inherited the Hebraic view of the world through the bible. But the totality of the tradition has not been passed on, only a selected series of ideas, buttressed by supporting texts. Later these texts were commented on by theologians in the light of their own agendas. The Genesis account of creation, as Clarence Glacken has pointed out, is noted for its brevity, [32] yet the text has spawned an enormous volume of commentary that has shaped Judeo-Christian attitudes more than the actual text. There are two accounts of creation in Genesis. The older narrative account (Genesis 2:4–25) shows God at work like a human artisan; the whole of creation is viewed as existing for humankind. The priestly account (Genesis 1) emphasises

---

30. Berry, Thomas, *The Dream of the Earth*, Sierra Club Books, San Francisco, 1988. Early in 1991 an entire *Insights* programme on Radio National was devoted to Thomas Berry's ideas. A tape is available from ABC Radio Tapes.
31. Both are available in Australia through the Canterbury Press in Melbourne.
32. Glacken, Clarence J., *Traces on the Rhodian Shore: Nature and Culture in Western Thought from Ancient Times to the End of the 18th Century*, University of California, Berkeley, 1967, p. 150.

magnificence: God brings forth the richness and beauty of the universe by his creative word, culminating in the creation of humankind in God's image.

A key text from the priestly account that has constantly been interpreted and re-interpreted is Genesis 1:26–28:

And God said: Let us make man in our image, after our own likeness . . . So God created man in his own image . . . and God blessed them and said to them: Be fruitful and multiply and fill the earth and subdue it . . . and have dominion.

In this limited and unfortunate text, which so vividly illustrates Israel's own antagonistic relationship to its difficult desert environment, we have a prime source and an apparent justification for contemporary ideas of exploitative development and anthropocentrism. For example, an editorial in *The Australian* recently used this text as a justification for the continued logging of Australian rainforests. There is, however, a deeper theological notion that can be drawn from the Genesis account of creation. Genesis makes it clear that human beings have dominion over the earth precisely because they are the icons or images of God. If read literally, Genesis says that only human persons, and more specifically, only men, are the image of God. The biblical preoccupation with idolatory leads it to deny any sense of divine iconography to the rest of creation. There is no recognition that the animals, birds, plants and the earth itself also mirror the creator. There is also a denial of any sense of consciousness and responsibility to the rest of creation; the cosmos and all within it are seen as mute realities there to be subjected to the needs of humankind. Adam's naming of the animals indicates that he defines their essential nature (Genesis 2:19–20). Their very existence is defined in relationship to him. Again, I emphasise that this strand of the Hebrew tradition does not exhaust the biblical view of the cosmos. The more positive aspect of the tradition is to be found especially in the Psalms and the Wisdom literature. But it is only the Genesis account which has been taken up by western culture to support the myth of development.

Implicit in the biblical account of creation is the presupposition that there is a separation between humankind and nature. Where does this distinction originate? Does it come from the biblical teaching that humans alone are created in the image of God, or does it come from primordial times when our ancestors were first able to assert their will over some animals through domestication? It is probably impossible to say but some of issues involved can be sorted out.

The first two chapters of Genesis describe a world which has been created for humankind and within which all life is arranged hierarchically, with man (who rules over woman) at the apex. Implicit in the text is a longing for purpose and order; God has designed the earth and divine power is revealed in the order of nature. Humankind, as God's agent, rules over the rest of creation. The view of God in these chapters is anthropocentric: God is seen in personal terms. And it is only human beings who can share in that personhood. Hence humankind's unique relationship with God which the rest of creation lacks. There is, however, a contrasting current of Christian mysticism that imagines God in terms that are more impersonal than personal. God is referred to by some mystics as a 'presence' rather than as a person. Some parts of contemporary spirituality have lost this awareness and are riddled with excessive emphasis on the personal nature of God. For instance, in the liturgical texts of the church, there is constant reference to God as 'Father'. The Latin word underlying this is so much more impersonal, referring to God as *Deus* (God). 'Father' is not the meaning of the word at all. Such usage illustrates the fear that some people have of more impersonal images of God.

No biblical text, even complete narratives such as the creation account, can be taken in isolation from the rest of biblical revelation, and biblical revelation itself cannot be divorced from the church's tradition, without the danger of introducing an element of distortion. The bible is not a total realisation of revelation; it must always be read in the light of tradition, the Christian community's ongoing interpretation of its history, belief and experience. Of course, the opposite is true too. Scripture acts as a norm for and check to the church's tradition. For instance, the doctrine of the redemption is generally presented in both theological writing and popular religion as being limited to the salvation of humans. The rest of creation is forgotten; it is outside the pale of salvation. There is another approach in the New Testament, however, which extends redemption to the whole of creation, represented by texts such as Romans 8:18–25:

For the creation waits with eager longing for the revealing of the sons of God; for the creation was subjected to futility, not of its own will but by the will of him who subjected it in hope; because the creation itself will be set free from its bondage to decay and obtain the glorious liberty of the children of God. We know that the whole creation has been groaning in travail together until now; and not only creation, but we ourselves, who have the first fruits of the Spirit, groan inwardly as we wait for adoption as sons, the redemption of our bodies.

While it is true that Pauline theology is 'much more soteriological that ecological' — in other words, Paul is more concerned with redemption from sin and the process of salvation, than he is interested in the environment — texts like the one above seem to link together the destiny of both humankind and creation. [33] Paul says (in the translation of the *New English Bible*) that 'the whole created universe groans in all its parts as if in the pangs of child birth' (Romans 8:22). Like the sons and daughters of God, the entire cosmos waits and longs for the liberation and fulfillment that comes with the Spirit of God. For Paul, creation is not an inert, unimportant backdrop to the human drama; it actually participates in the same process of liberation and salvation as do human beings.

The distorted pattern drawn by later theology from the biblical description of creation was further twisted as the early Christian church struggled to explain itself to the Roman world of the 2nd, 3rd and 4th centuries. Genuine Christianity has never been fundamentalist and, as it spread, it searched for ways to explain itself in terms that made sense to the people of the time. The patristic theologians realised that they could not simply say 'Believe this because the bible, or Christ, or whoever, says so'! Intelligent and civilised people sought grounds for their belief. The philosophy that Christianity adopted to achieve this was the late Roman adaptation of Plato's teaching — neo-platonism. One of the church's models for its attempt to convey faith in terms of local culture was the Jewish theologian, Philo of Alexandria (c.30BC–50AD), who had attempted to integrate the Hebrew bible with Platonism. Philo was influential in the work of Clement of Alexandria (c.150–215) and Origen (185–254), both of whom were associated with the catechetical school of Alexandria, the first institution ever developed for the training of converts to Christianity.

But in adopting the philosophy of neo-platonism, Christianity imported new concepts into theology which over-emphasised particular elements, leading to distorted views. An example of this is the importation of the dichotomy between soul and body which

---

33. Carmody, John, *Ecology and Religion: Towards a New Christian Theology of Nature*, Paulist Press, New York, 1983, p. 92.

departed from the more integrated Hebrew and biblical concept of the person. We should note here that neo-platonism was not the product of a flourishing culture, but of a society in decline — the late Roman world. Rosemary Radford Ruether described this world thus:

Late antique culture was obsessed with the fear of mortality, of corruptability. To be born in the flesh is to be subject to change, which is a devolution toward decay and death. Only by extricating the mind from matter by ascetic practices, aimed at severing the connections of mind and body, can one prepare for the salvific escape out of the realms of corruptability to eternal spiritual life. All that sustains physical life — sex, eating, reproduction, even sleep — comes to be seen as sustaining the realm of death, against which a mental realm of consciousness has been extracted as the realm of 'true life'. [34]

Neo-platonism held that the fullness of life was in the soul, which had been captured, dragged down and degraded by the body. Asceticism was seen as the only way of escape from both the body and the world. The world was the dwelling place of the demonic. Those who sought to understand nature did so by making a pact with the devil.

Through patristic theology, platonised concepts became fixed in Christian theology and through theology permeated the intellectual élites of western culture. The dichotomy created by platonism between soul and body, matter and spirit, God and the cosmos, has caused enormous tension in western Christian culture by making us suspicious of matter and of all that is not spirit.

In the Middle Ages there were some who broke out of this destructive mind-set. Saint Francis of Assisi, Saint Hildegarde of Bingen, Master Eckhart and many of the medieval and post-reformation mystics freed themselves from this intellectualised anthropocentrism. But the idea of the earth as a 'vale of tears' still predominates in the tradition: the church's dream is not so much of this world, but of a better world to come.

With the 16th century Renaissance and the scientific revolution of the following centuries, the world gradually disentangled itself from the church. Nature, previously seen as the dwelling place of the demonic, was gradually secularised, governed by laws that could be identified and predicted. With the vast technological changes of

---

34. Ruether, Rosemary Radford, *Sexism and God-Talk: Towards a Feminist Theology*, 1983, pp. 79–80.

the 18th and 19th centuries, mechanistic concepts developed which saw nature as a vast store-house of useable material that could be exploited. The romanticism of the 19th century attempt by alienated artists to halt the progress of machine culture — but only a temporary one. Industrial society was underpinned by the myth of progress: the processes of history were to carry humankind onward to an ever fuller life based on the (presumed) superabundance of the earth.

We are now at the end of western industrialisation. It is a tragedy that third world countries are rushing into the same mistakes as those made in the 19th and early 20th centuries by the west. The myth of development has already turned much of the world into a wasteland. As science leaves the old world of industrialisation behind, it is developing a new cosmology that is radically different from the Christianised Greek cosmology of the church. Contemporary science and the contemporary church emerge from totally different understandings of the nature of reality. But, as I said at the beginning of this section, a pontoon has been thrown across the abyss: concern about the environment is creating space for dialogue between science and the church.

## The growth in population

All the environmental problems have been compounded by the extraordinary contemporary growth in the human population. There are those who argue that the earth can carry many more people, but is this true? And can it carry such numbers without destroying the world as we know it? The problem was clearly expressed recently by Prince Philip Duke of Edinburgh, President of the Worldwide Fund for Nature:

I think that it should be glaringly obvious that as the number of people increases, so does the damage to the global biological and physical systems. These are our life-support systems and we damage them at our peril. All the evidence points to the conclusion that the human species has offended against one of nature's basic laws. We have over-stretched the carrying capacity of our habitat. [35]

Unless we want to live like ants in an anthill, with everything sacrificed to the needs of survival, there has to be a cap on population.

---

35. The Duke of Edinburgh, quoted in *The New Road: The Bulletin of the WWF Network on Conservation and Religion*, no. 16, Oct-Dec. 1990.

Many people today talk about 'concessions to the greenies' as though the preservation of ecological systems were some type of luxury. If the earth were reduced to anthill status, we would no longer be human beings in any real sense. Without the other living plants and animals, without natural beauty, our humanity would be so debased as to destroy civilization, art and spirituality.

The world's population is increasing exponentially:

In 1575, after a century of rapid growth had added 100m to the total, the world's population reached 500 million. By 1825 it had doubled to a billion, by 1925 it was nearly 2 billion, by 1975 only a fraction under 4 billion. Note how the time to double dropped from 250 years to 100 years and then to 50 years. If, as seems likely, it remains at 50 years for the next phase of the cycle, there will be nearly 8 billion people on the earth's surface by 2025. [36]

Historical demographers, like McEvedy and Jones, argue that after 2025, the population rate 'must slow down'. They argue that population will never reach the earth's theoretical limit of around 20 billion people, that it will level out in the 21st century at between 8 and 9 billion. They claim that this will be achieved by higher living standards and the spread of education — although it is not clear what they mean by 'higher standards of living'. Other evidence also shows that an essential element in population control is education and the liberation of women. There is a natural link between the control of fertility and the control of one's personal life through knowledge and freedom. It is significant that fertility is highest in countries where women are still oppressed and uneducated. McEvedy and Jones state the alternative bluntly: 'If population doesn't slow down spontaneously it will have to be stopped by some sort of catastrophe, either man-made, microbial or nutritative'. Certainly, population rates are beginning to slow. But because of very high birth rates in less developed countries, the world contains a large proportion of women of child-bearing age. In 1989 thirty-seven per cent of the third world population was under fifteen and therefore had still not reached the prime reproductive period. A small saving grace is the fact that a child born in a poor nation will use far fewer resources than one born in a developed country.

In the light of all this, the situation in Australia seems idyllic. We have such a vast land and so small a population. On the face of it,

---

36. McEvedy, Colin & Jones, Richard, *Atlas of World Population History*, Penguin, Harmonworth, 1978, p. 350.

Australia ought to take many more people. It was asserted recently by the Sri Lankan priest, Father Tissa Balasuriya, on the ABC Radio programme *Insights* that Australia could absorb 150 million Asian peasant farmers! Allowing that this was a massive exaggeration aimed at getting Australians to face their responsibilities to those in need, it nevertheless highlights a common assumption that this vast country could carry many more people. The situation is not so simple. Most of Australia is desert or marginal land, incapable of being farmed without massive environmental destruction. Much marginal land has already been destroyed. Significant land degradation and soil loss means that we are able to produce less from the same land. Deforestation and irrigation has led to large areas suffering from salination. The Riverina and parts of northern Victoria near the Murray, as well as large tracts of the wheat lands of Western Australia, are already badly affected by salination. Dr Tim Flannery of the Australian Museum in Sydney argues that Australia's ecology can sustain a population of six to twelve million: we already have seventeen million.

Australia has the fastest growing population in the western world. While most sensible Australians are happy to support such admirable activities as recycling and the preservation of wilderness, many people have cultural, psychological or religious problems about curbing the number of their offspring. The urge to reproduce is difficult to frustrate; human psychology is geared to it because of many centuries of low population and high infant mortality. Further, in today's society to have a child is seen as a very personal and private matter. Increasingly however it is becoming an issue with major social implications. To have a child in Australia puts much more pressure on the world ecosystem in terms of energy use, than does having a child in a poor country. It is easy to label people concerned about the environment as being 'anti-child', but we have to ask what type of world we will leave to our children and grandchildren if we do not face these hard population issues. The Australian church simply cannot side-step the moral and ethical dilemmas embedded in the population question.

## Immigration

Also involved here is the question of immigration. A recent article by Dominican Father Anthony Fisher asserts that 'few [Australians] would be self-critically aware of the ideologies that inform their position on immigration'.[37] Among those apparently unaware of

---

37. Fisher, Anthony, 'The Immigration-isms', *Uniya*, Autumn 1991, pp. 4–5.

their presuppositions are 'fringe zero population growth groups and some environmentalists'. Fisher argues that population is not a threat to the environment and that environmental issues must be tackled from a global perspective rather than from a 'selfish and inward looking isolationism' which quarantines us 'from sharing our "national park" (with) . . . anyone else'. He challenges the idea of separate nationalism and argues for increased immigration on the basis of international social justice.

Australia has the lowest population density of any major country, even after discounting its deserts, and is well-off by world standards, even in times of recession. On this basis generous immigration quotas — more generous than our present levels — would be justified and morally required. [38]

I would be the first to admit that it is difficult to oppose some of these arguments, especially when they are placed within an international social justice context. The problem of criticising the migration intake is compounded by the fact that so many Australians are born overseas, or are the children of migrants; family reunion is important to them. Also the pressing world problem of refugees cannot be ignored; the Australian Catholic church has had a laudable commitment to them. However, I think there is a serious flaw in the strong pro-immigration argument propounded by Fisher. He seems to have ignored the evidence that an increasing Australian population, especially one living at present consumption levels (and there is no evidence that this is going to change quickly), simply has to place more stress on the production capacity of the country. And this production capacity has a direct relationship to the environment, for more marginal land will have to be brought into production and there will have to be significantly greater clearance of the tree cover. An increase in population of one million in Australia (our population has actually grown by one million over the last five years) is estimated by Flannery and Telford to equate to adding sixty million people to the population of a poor country, because Australians use sixty times more energy. Even if Flannery and Telford's figures are not precisely accurate, their point makes eminent sense. [39] To bring more people to Australia is to increase the demand on the country's capacity to produce and thus to cause environmental degradation.

---

38. Fisher, art. cit., p. 5.
39. Flannery, Tim & Conlon, Telford, 'A Vaccine for the Plague', *Australian Natural History*, 23/2(1989), p. 155.

There is also a bias lurking in Fisher's apparently strong ethical argument. In the long-term, it may be morally more justifable to preserve Australia as a world 'national park' for the future. There is agreement to preserve Antartica as a world heritage area, so we ought be thinking of Australia in the same way. I would argue that we have a profound moral obligation to preserve what is left of wilderness for coming generations. One way to do this is to keep a cap on population. Moral responsibility is not just a numbers game, as Fisher seems to suggest; it must also take into account the question of quality of life for coming generations.

For a long time the church simply assumed that immigration was a good thing. As we face the 21st century, this can be no longer taken for granted. Certainly, we should make exceptions for genuine refugees, especially those with whom Australia has a historical connection, such as the Vietnamese. However, it is hard to see why we should take immigrants for whom we have no historical responsibility. Our ability to absorb new people in Australia is not unlimited. We must ask here what the Catholic church in Australia can contribute to this issue that is not provided by anyone else.

## Towards a new cosmology

The thought of Thomas Berry can help us here. I said earlier that the central philosophical and theological problem facing us is a cosmological one. Contemporary Christianity is working out of a world view that is out of date. Berry is one who has begun the construction of a new cosmology. Theology, he says, must begin to take the world seriously and recognise the place of humanity in creation. Human history is very brief compared to biological history. The processes of the earth are four billion years old while human beings have been on earth for less than 60 000 years. The starting point of the new cosmology should not be the usual sources of theology — the bible, the church, tradition, doctrine, and certainly not the pope or the hierarchy — but the world. Historically and factually, the universe is the primary given.

The first element in the new cosmology is a larger concept of time. The cosmologies of Aristotle, Plato, Augustine, Aquinas and Bishop Jacques-Bénigne Bousset (whose ideas about the history of the world were normative for the period of the 17th to 19th centuries) were limited in space and time. While Aquinas had developed a philosophical concept of a possible infinite time, his historical ideas — like those of the other great minds of the Christian world — were limited by biblical revelation. Bousset, who saw the church as a non-historical and unchanging reality, had even worked out the 'exact' age of the universe by computing the years mentioned in the bible! For our forebears, then, historical time was circumscribed by biblical history. It was a conceivable and manageable reality. The idea of eternity, defined as the ever-present totality of all time, was also conceivable for them. They thought of it as that static point outside time on which God stood, viewing and directing the events of history. The idea of eternity, though, only makes sense if historical time is circumscribed by particular events such as the biblical story of creation and 'the day of the Lord' at the end of the world.

Today's theology cannot begin with the parameters of biblical history, for this is too narrow. We have to begin with biological history which has gone on for at least four billion years — a time we are simply unable to imagine, let alone conceptualise. Theology has to move to a concept of God who is *part of* time not apart from it, whose creativity is manifested through the evolution of biological processes and through the history of the cosmos. The universe is not thought of as a reality that was created suddenly; it is an ongoing energy event, a self-emergent dynamic process that is caught up in its own inner process. It is self-explanatory; we do not need God to explain it. Theological effort today must therefore be directed towards discovering God *in the process*.

Another major element in developing a new cosmology emerges from the way in which contemporary science conceives of life. It sees life as an interactive continuum from primitive forms to the most highly evolved, the human person. It is the genes which pass on this ever-increasing complexity of life; all reality is thus related genetically. Berry argues that humankind needs to discover its genetic roots, our 'genetic coding', in order to appreciate our rootedness in the whole cosmic process. Healthy people do not usually destroy the very substance of which their life is a part. Thus the rediscovery of the profound genetic inter-relatedness between ourselves and everything else in the universe strengthens the demand that we work to conserve life, rather than destroy it by attempting to 'develop' it.   In this context, it is significant to note that Christian theology often has a limited definition of 'life'. When addressing texts like 'I came that they may have life and have it to the full' (John 7:10), the traditional interpretation is in terms of human life and today especially in terms of a full psychological development which finds its fulfillment in a spiritual life. Biological life is simply neglected. But it is our genetic coding that links us into the immense web of life. Thus when Jesus in John's gospel speaks of fulness of life this can be interpreted as also including the whole of biological life. The cosmos is also differentiated, that is, it is made up of individual energy constellations. The universe is a collection of individual subjects in interactive communion with each other, all linked by their genetic coding. While humankind is the highest form of active subject, what should be stressed is our part in the communion of life, rather than our separation from it.

Creation myths provide all societies with their most important cultural underpinning, and from them comes the basic constitutive element of all theologies. As we have already seen, for example, the Genesis account provides Judeo-Christianity with a definition of humankind's relationship to God and to the rest of nature. It also explains the origin of sin and evil and the ambivalent attitude of human beings to their lot; we still want to be 'like gods'. Thomas Berry points out that because creation myths define cultures, they also divide cultures from each other. As long as each tradition insists on its own scriptures and myths, they emphasise their differences. Science, then, can provide us with the basis for ecumenism in the widest sense. The scientific myth gives us all a common story of origins, a story that begins with the primal atoms and moves through the whole evolution of life. It is precisely this type of common myth we need as we face problems that are not just local but worldwide.

Talk like this, of course, creates problems for some Christians. These problems constellate around fears of nature worship and pantheism. It is true that environmentalism began outside the context of the mainstream churches and it is true that some environmentalists have begun, consciously or unconsciously, the construction of a kind of environmental religion. John K. Williams in a paper entitled 'The Religion of Environmentalism', states that 'the extreme environmentalist' worldview is basically religious and must therefore be rejected by Christians. [40] He argues that extreme environmentalists have a static view of nature which means that nothing in nature can be changed by human intervention. Thus all economic development is excluded. They have 'an uncritical acceptance of the snapshot view of nature . . . what is shown to obtain at a given moment of time is set up as desirable, inviolable — to all intents

---

40. Williams, John K., 'The Religion of Environmentalism' in the Australian Institute for Public Policy series *Economic Witness*, 6 July 1990.

and purposes, as sacred and to be worshipped'. He points out that nature is ever-changing, prodigally wasteful and quite brutal in its elimination of weaker species and individuals. He accuses 'extreme environmentalists' of having 'a romanticised, totally unscientific view of "Nature" and of humanity's place in nature (which can be largely traced back to the writings of the unspeakable Rousseau)'.

Ignoring the gratuitous comment about Rousseau, it seems to me that Williams has merely set up a straw target and demolished it. Certainly there are people who see environmentalism as a form of religion — and it is a far superior form of religion to the worship of greed espoused by many large corporations — but Williams produces no names of 'extreme environmentalists' who have a sacred 'snap shot' view of nature. The reason for this is that there are no serious environmentalists with such an unsophisticated view of reality. The only names he mentions in his paper are Paul Ehrlich and David Suzuki, neither of whom has a 'sacred' view of the cosmos.

The more significant argument against a religious understanding of environmentalism is the accusation of pantheism. It is significant that the last person burned by the Roman Inquisition in 1600 was the pantheist Dominican, Giordano Bruno. The essence of the theory of pantheism is that nature *is* God. However, what is being argued by Christian environmentalists such as Charles Birch is quite different. They are arguing that God can be found *in* nature, that the created world speaks of God. This is not pantheism but panentheism. It is a perfectly orthodox position for any Catholic (or any Christian for that matter) to hold.

## To save the earth

Unless the church faces up to the environmental issue, it will have little to say to contemporary culture. The need to broaden its concept of life is the greatest challenge facing contemporary Catholic Christianity. In a way, it is really only the religious and spiritual traditions that can provide the philosophical and theological foundation that contemporary environmentalism requires. That is why I see it as the most serious issue we face within the Australian Catholic church today. We have the future of the whole of our country in our hands. Australia is a unique bioregion and it is one of the few wilderness places that is not immediately threatened by overpopulation. Thomas Berry uses this term bioregion to describe 'an identifiable geographical area of interacting life systems that is relatively self-sustaining in the ever-renewing processes of nature'.

We are faced in Australia with serious and pressing ethical, philosophical and theological problems. Why preserve a rainforest? Is it not just x cubic metres of newsprint or toilet paper? How do you talk about the value of the living creatures within it? How do you quantify its beauty, its aesthetic value? These are profoundly religious and spiritual questions that the Catholic community cannot avoid confronting without being totally untrue to itself and to its ethical tradition. Australian Catholics cannot shift the responsibility for addressing these questions to the pope or to overseas theologians or to whoever. They are peculiarly *Australian* questions and we have to address them. In the process of answering these questions the Catholic community will begin to act against the truly demonic forces that are operative within our society, with their fixation on the concept of economic 'development'. The stand has to be made now, otherwise our Australian bioregion will most certainly be destroyed.

These questions have already permeated to the popular level. Ordinary Australians are concerned about the future of the planet and specifically about the future of our own environment. If the local Catholic church does not see this issue as central, it will neglect the most important moral and theological issue facing Australian society and thus render itself irrelevant to a major part of its own constituency.

This danger of irrelevance is a real one for the Australian Catholic church. There are strong forces at work within the Catholic community that will, unconsciously but nevertheless decisively, turn the church back in on itself, with the real possibility of a retreat to a sub-culture. That is why the environmental question is central. It demands that the church turn outward to important questions in the world, questions like the relationship of humankind to the cosmos, the reconciliation of science and religion, the nature of life itself and its intimate interconnectedness. Environmentalism is also a way of focusing on social justice and population issues. At heart, the environmental crisis forces the Catholic community to rethink a number of its basic presuppositions about God, the world and humankind's place in it. No other religious community is better placed to do this.

# CONCLUSION

The environment issue is one of those basic questions that niggles away in the mind and conjures up other basic issues such as: why do I remain a Catholic? This question emerges from the legitimate problems that many Catholics have with the church. It seems such a self-interested institution, constantly concerned about its own survival. Jesus, as well as the individual Catholic, seems to be lost in the whole ecclesiastical apparatus. Many feel that the Vatican has become an albatross around the neck of the Catholic community. It seems intent on maintaining its own power, often at the expense of those people who are most committed to the church. Many critics point to the inconsistency of talk about social justice when there is considerable internal injustice in the church against women, the laity (who have almost no say in what happens in the church), priests who have left the active ministry, and even against some of the most pastoral bishops in the Catholic world. This list could go on and on. So the question cannot be avoided: why stay?

I think that most committed Catholics have discovered that there are really two simple answers to this question. The first is that for many of us there is nowhere else to go. Nothing else will satisfy us; this is our home and we are staying. And we are certainly not going to hand over the church to the reactionaries by leaving it. As Saint Peter said so clearly in John's gospel: 'Lord, to whom shall we go? You have the words of eternal life and we have believed' (John 6:68–69). Peter addressed this remark to Jesus and not directly to the church, but belief in Jesus implies membership of the community of his followers. How can the giant structure of Roman Catholicism be the 'little flock' of Jesus? It can certainly make a genuine historical claim to be part of it — in the old days, Catholics would have said that they were *all* of it, but now we are more humble!

Membership of the Catholic church also roots us in the institution which helped create the culture in which, as westerners, we were born and to which we belong. It places us in the tradition of a whole series of great men and women in both a human and spiritual sense. It puts us in contact with an extraordinary intellectual and theological tradition that has created the most consistent attempt to understand the meaning of human existence in the entire history of humanity. Only fools would abandon all this. Certainly there are those who have been hurt deeply by the church but it is tragic when personal hurt becomes so ingrained that a truly objective perspective is lost.

The second answer to the question is equally impelling and in a way brings us back to the environmental question. Catholic Christianity is one of the few religious expressions that has built into its belief system a dynamic that is essentially creative; it is not only a religion of book and law, but also one of living tradition. There are many people today and there have been many more in the past — who claim that we must return to the 'simple faith' of Jesus. This is not only impossible, it is also a serious mistake. History cannot be undone; the contemporary world has come a long way from Jesus. There is no way that we can leap over our history to resurrect the past.

Membership of the Catholic church places us in one of the few truly creative institutions in the world today. In *Mixed Blessings* one of the major themes I developed was the sheer creativity of world Catholicism. The Catholic church has already penetrated the emerging cultures of the future in the third world. More than sixty-five per cent of all Catholics are now from Africa, Latin America, Asia and the Pacific. The tremendous dislocation — or 'mutation' as I have called it — that Catholicism has experienced since the Second Vatican Council has opened the church up to new ideas and new realities, in a way unrivalled in its entire history. There is no doubt that world Catholicism has a future. And an integral part of that future is linked to environmental questions. The creativity of the church must be primarily directed to the maintenance of the life systems of the earth. To ignore creation is to ignore God.

I have to admit, however, that I am less certain about the Australian brand of Catholicism. Certainly, many of the same choices and challenges confront Australian Catholics as confront world Catholics. We are not called to resurrect the past, but to live in the present and work toward the future. This is a massive task. I have indicated through this book there are elements at work within Australian Catholicism that could easily turn us inward and away from the real issues. The tragedy is that the Catholic church may never actually face these issues because of the self-enclosed fear of its bishops, the personal confusion of its clergy, the voiceless impotence of its laity within the structures of the church, and the short-sightedness and lack of any clear agenda that is so characteristic of Australians and that results in an absence of any coherent thought about the future. Certainly, as this book indicates, many creative things have been achieved, the most important of which is the extraordinary renewal of the Catholic laity. But structural changes have to occur to incorporate that renewal, and it is at this point that hierarchical fear and self-interest and a lack of opportunity for lay initiative may cause a tragic breakdown. The blandishments of the reactionaries toward a safe and comfortable sub-culture are also a temptation, although the realism typical of most Australians will probably save Catholicism from this form of escapism. The biggest danger that I see is the lack of any coherent plan. It is so easy to stumble into the future with no set agenda.

One thing is absolutely certain: Australian Catholicism does have to make a choice. The options also are clear: to be true to its creative tradition and thus contribute to the formation of the new world, or to retire into a traditionalist rump. Either way, the choice will be decisive for a number of generations to come.

# INDEX